The Naked Android

The Naked Android: Synthetic Socialness and the Human Gaze illuminates the connection between the stories people tell, their expectations of what a robot is, and how these beliefs and values manifest in how real robots are designed and used.

The introduction of the "human gaze" articulates how peoples' expectations and perceptions about robots are ultimately based on deeply personal cultural interpretations of what is artificial or human and what problems social robots should—or should not—solve. *The Naked Android* clarifies how human qualities like understanding and desire are designed into robots as mediums as well as projected onto them by the people who live with them.

By investigating the fluidity of identities across human culture and social robotics, this book unpacks the contextual complexities of their interactions and mutual influences. Using ethnographic methods including in-depth interviews with a variety of stakeholders, each chapter explores how people are designing social robots, the experience of living with robots, and people whose jobs it is to dream about a future integrated with robots.

Key Features:

- Introduces the concept of the "human gaze" (and the "robot gaze") as means of understanding how people live with robots.

- Each chapter includes in-depth interviews with people who make, live with, or create art about robots.

- Using ethnographic methods, paints a vivid description of the interconnecting influences of science fiction, human imagination, and real technology.

Julie Carpenter, PhD, is a social scientist who explores human behaviors with emerging technologies, often forms of AI. A great deal of her research has focused on human attachment to robots and other forms of artificial intelligence. She is an external research fellow in the Ethics + Emerging Sciences Group at California Polytechnic State University and has held this role since 2015.

Chapman & Hall/CRC Artificial Intelligence and Robotics Series
Series Editor: Roman Yampolskiy

Transcending Imagination: Artificial Intelligence and the Future of Creativity
Alexander Manu

Responsible Use of AI in Military Systems
Jan Maarten Schraagen

AI iQ for a Human-Focused Future
Strategy, Talent, and Culture
Seth Dobrin

Federated Learning
Unlocking the Power of Collaborative Intelligence
Edited by M. Irfan Uddin and Wali Khan Mashwan

Designing Interactions with Robots
Methods and Perspectives
Edited by Maria Luce Lupetti, Cristina Zaga, Nazli Cila, Selma Šabanović, and Malte F. Jung

The Naked Android
Synthetic Socialness and the Human Gaze
Julie Carpenter

For more information about this series please visit: www.routledge.com/Chapman—HallCRC-Artificial-Intelligence-and-Robotics-Series/book-series/ARTILRO

The Naked Android

Synthetic Socialness and the Human Gaze

Julie Carpenter, PhD

CRC Press
Taylor & Francis Group
Boca Raton London New York

CRC Press is an imprint of the
Taylor & Francis Group, an **informa** business

A CHAPMAN & HALL BOOK

First edition published 2025
by CRC Press
2385 NW Executive Center Drive, Suite 320, Boca Raton FL 33431

and by CRC Press
4 Park Square, Milton Park, Abingdon, Oxon, OX14 4RN

CRC Press is an imprint of Taylor & Francis Group, LLC

© 2025 Julie Carpenter

Library of Congress Cataloging-in-Publication Data
Names: Carpenter, Julie (Julie G.), 1969– author.
Title: The naked android : synthetic socialness and the human gaze / Julie Carpenter.
Description: First edition. | Boca Raton, FL : Taylor and Francis, 2025. |
    Series: Chapman & Hall/CRC artificial intelligence and robotics |
    Includes bibliographical references and index.
Identifiers: LCCN 2024029138 (print) | LCCN 2024029139 (ebook) | ISBN 9780367772543
    (hardback) | ISBN 9780367772529 (paperback) | ISBN 9781003170457 (ebook)
Subjects: LCSH: Robots—Social aspects. | Human-robot interaction. |
    Human-machine systems. | Sex dolls | Robotics—Moral and ethical aspects.
Classification: LCC TJ211.49 .C36 2024 (print) | LCC TJ211.49 (ebook) | DDC 629.8/924019—
    dc23/eng/20240822
LC record available at https://lccn.loc.gov/2024029138
LC ebook record available at https://lccn.loc.gov/2024029139

ISBN: 978-0-367-77254-3 (hbk)
ISBN: 978-0-367-77252-9 (pbk)
ISBN: 978-1-003-17045-7 (ebk)

DOI: 10.1201/9781003170457

Typeset in Minion
by Apex CoVantage, LLC

I count myself incredibly fortunate to have not one, but two good friends in Deborah Fritsch and Joan M. Davis. This book is dedicated to them with gratitude and affection.

Contents

Preface

BETTER, WORSE, OR THE SAME?

You are seated in a darkened room, leaning into a cold, mask-like machine pressed against your face while a clinician uses their expertise to assess your eyesight. The professional manipulates the various lenses stacked within it, each *click* of the lenses sharpening or blurring the letters and shapes projected in front of you. With each change, the clinician prompts you, "Better, worse, or the same?" The outcome of your series of replies and based on how well you can accurately read what is projected pinpoints your vision for the clinician, who then labels it with numbers that have meaning to other optical experts. One result of this human-machine experience is that you may eventually receive new glasses or contact lenses, tools worn on your body to correct vision to a widely agreed-upon clinical standard of what is considered *normal*.

BETTER, WORSE, OR THE SAME?

This ordinary experience of vision testing is a metaphor for how I often think of the extraordinary directions that people suggest we use AI and robots. In some ways, my overarching research questions over the last decade often boil down to hundreds of branching variations of, "Does this social human-robot interaction scenario make peoples' lives better, worse, or the same?"

My first thoughts are about human safety concerns, which range from data privacy handling to users' physical safety when interacting with embodied AI, like robots. When a new robot is released into the general population, or *into the wild*, what risk and safety factors have been prepared for when people encounter unanticipated situations or react spontaneously toward robots not designed for social of interaction? For the former, I immediately think *why design for socialness? What problem(s) does that developer think socialness will solve, and for whom? What considerations, if any, have been made for the user's mental and emotional safety when designing for socialness?*

The latter issue regarding spontaneous socialness additionally raises these initial questions: *what is the situation in which these interactions take place? What factors do the users identify as leveraging a social aspect to their interactions with the robot? What benefits and harms may emerge for the user by constructing this narrative?*

As a researcher with a focus on human-robot interaction and the cultural aspects of technology adoption, I have dedicated much of my research to exploring the evolving relationship between humans and robots. As these machines become increasingly sophisticated, integrated into our daily lives—and normalized—it is crucial that we understand not only the technical aspects of robotics but also the complex emotional and ethical dimensions of our interactions with them. In this book, I aim to provide an interdisciplinary examination of a concept I call the *human gaze*, and its significance to ethical considerations in all aspects of robot research and development.

Here, I introduce the concept of the *human gaze* and its implications for how to think about robot research and development, examining personal narratives, exploring potential consequences and outcomes, and discussing the future of human-robot interactions. It is my hope that this work will serve as a valuable resource for curious minds of all sorts—including researchers, robot makers, and practitioners from a wide range of disciplines—and foster interdisciplinary dialogue and collaboration in the pursuit of ethical, responsible, and sustainable robotics.

This book is meant to be a valuable means of exploring and describing emerging cultural beliefs, providing insights into the complex interplay between individual experiences and the evolving cultural landscape. By adopting a phenomenological framework and employing ethnographic methods, as researchers we can gain a deep and nuanced understanding of the ways in which personal narratives both reflect and shape emerging cultural beliefs about how we interact socially with robots. This approach contributes to the ongoing dialogue about the dynamic nature of culture, AI, and the factors that inform human experiences and understandings of the world with AI and robots in it.

The field of human-robot interaction possesses the remarkable capability to shed light on aspects of human nature that might otherwise remain concealed. It captivates my interest due to its ability to reveal the threads that connect us, uncovering the intricacies of this abstract concept called *culture.* In the case of this book, the idea of AI and social robots as ways of understanding who we are as humans was the genesis of the phrase *the naked android.*

Certainly, the word *naked* is a reference to our human vulnerabilities, including hubris. Famously, Derrida connected the concepts of both *nudity* and *shame* as distinctly human qualities. In *The animal that I am (More to follow)*, he tells a story about his cat observing him while he is naked, which prompts him to explore the concept of shame. This experience leads him to question conventional boundaries drawn between humans and animals. Derrida posits that the shame humans feel when naked in the presence of animals reveals the deep-rooted anthropocentric distinctions we create between ourselves and other species, arguing that this moment of nakedness symbolizes a defenselessness, vulnerability, and a shared existence with animals, challenging the clear separations between humans and animals traditionally upheld and agreed-upon culturally. He wrote,

> It is generally thought, although none of the philosophers I am about to examine actually mention it, that the property unique to animals and what in the final analysis distinguishes them from man, is their being naked without knowing it.

> (DERRIDA, 2002, P. 373)

He continues that animals "without knowledge of their nudity" are "without consciousness of good and evil" (*ibid.*). Derrida's quote is an invitation to reconsider the differences between humans and animals, and it also speaks to a broader idea of deconstructing established social categories and encouraging deeper philosophical inquiry.

Furthermore, I purposely use the term *android*, a word of many different cultural associations. Among the meanings of *android* is a humanlike robot that is perceived to be male gendered, and this book examines the first waves of social robots, how and when they are gendered, why we gender robots, and user expectations of robot gendering. The use of the word *gynoid* has historically been the language used for robots that are female gendered, yet popular use of the droid language has defaulted to the masculine term, even though a significant number of social robots are designed with a range of nongendered designs or designed to resemble women. Thus, this work challenges current perspectives around gendered and nongendered robots and asks us to (re)consider what expectations we have of social robots in relation to ourselves. When designing social robots in ways that represent gender, race, age, ethnicity, and religion, important questions arise about if these robots are maintaining sociocultural *status quos* instead of exploring innovative, perhaps unexplored options.

Like many disciplines within the social sciences, human-robot interaction carries a historical legacy of its own biased categorizations of groups of people and voyeuristic research traditions. In this way, too often research goals perpetuate endlessly repetitive and objectifying narratives about communities that the researchers remain detached from in any meaningful manner. Therefore, I find myself grappling with the question of how to genuinely comprehend, interpret, attentively listen to, and wholeheartedly acknowledge the observations and reflections stemming from people's lived experiences. In my work, I earnestly endeavor to place this reflection at the core of my scholarly pursuits.

Thank you for joining me on this journey, and I look forward to the conversations and insights that will undoubtedly arise from our collective exploration of the *human gaze* and its implications for the future of robotics.

Acknowledgments

Thank you to the CRC Press, who stood by my vision for this book through these long, strange times. An especially warm thanks to my editor, Randi Cohen-Slack. Your patience has shown me that editing, much like writing, is an act of empathy.

Cheers to the wonderful Solomon Pace-McCarrick, with whom I had the opportunity to work with as an endlessly resourceful and supportive editorial assistant early on this book. Solomon, you were the first keeper of my stories. Thank you for standing by me through the winding research process.

To Dr. Patrick Lin and the Ethics and Emerging Sciences Group at California Polytechnic State University, I owe a debt of gratitude for their enthusiastic support over the years.

Thank you Ronak K. Bhatia and Allison J. Youngdahl; your unique blend of humor, warmth, and loyalty have enriched my existence and at times been my shelter. And while the world may demand productivity, Jen Van Dijk and Amy Bensinger, your talents for dragging me into an abyss of amusing distractions are impressive, terrifying, and appreciated.

The engineering spirit and scientific curiosity of my brother Brian L. Wajdyk and the teaching legacies of Drs. John D. Bransford and William Hart-Davidson are in my thoughts. May their memories be a blessing.

I am thankful for the emotional support and intrusive interferences of my cats Ludo and Bonanza Jellybean, and their partner in crimes, The Chicken, and for their being the black hole of productivity in my life. Here's to never getting anything done together.

To write this sort of work is to engage in a form of ephemeral archaeology, digging through the layers of the past and present to unearth the bones of future realities. Thus, a very warm thank you to all the archivists, librarians, and media relations folks who helped me track down quotes, images, visual artifacts, and sources. Your dedication—to the stories, to the seekers of stories, and to the untold stories yet to find their voice—is a reminder of the power of collective memory and shared history.

A special acknowledgment for the body of work of science fiction author Zenna Henderson, whose stories about curiosity, memory, inter-generational trauma, empathy, and how we *other* reverberated with me as a child and continue to resonate with me decades later. Henderson's narratives are not just stories but simulations, scenarios that timelessly compel readers to confront ethical, societal, and existential dilemmas. Like all the best science fiction writers, Henderson's legacy is not just in the worlds she created but in the minds she inspires.

Thank you to those who allowed themselves to be interviewed for this book and have willingly stepped into the labyrinth of this endeavor; you have not only contributed to the expansion of knowledge but also engaged in the profound act of co-creating reality. Thank you, then, for this gift of perspective, for the ability to see beyond the surface sheen of the world as we each know it. Our journey together reaffirms that it is through our engagements, our choices, and our shared pursuit of understanding that we carve out the essence of what it means to be truly human.

CHAPTER 1

Welcome home, robots

INTRODUCTION

This is not a story about technology per se—it is a story about what happens to people when robots enter their lives.

In the introduction of social robots into society, there arises a new mode of existing together. The social robot does not merely inhabit the human world; it transforms it, forcing people to confront presuppositions about agency, autonomy, and the essence of the social bond itself.

When a robot assists an older person or acts as a companion to a child, the interactions, though artificial, are imbued with real social significance. These robots do not merely act; they participate in the creation of social reality, influencing both the individual and collective psyche. It is akin to the moment when a stranger enters a close-knit group: the dynamic shifts, roles are reevaluated, and the group's perception of itself and its purpose inevitably changes.

As social robots become ubiquitous, people may find their behavior subtly shifting. Conversations may be more guarded, or perhaps more open, depending on how the people feel about the robots. Social norms begin to warp slightly—what does politeness mean in response to a machine that simulates understanding? Each interaction with a robot becomes a stone in the mosaic of this new social reality.

The questions thus arise: how are the societal norms and roles observed being subtly rewritten by the presence of these synthetic social actors? What does it mean to be a being among other beings, human and non-human alike? How do these artificial entities influence perceptions of authenticity?

DOI: 10.1201/9781003170457-1

In grappling with these questions, it is imperative to recognize the transformative effects these artificial beings have on human interaction and societal dynamics. This shift prompts a deeper examination of the nature of consciousness and the essence of human social fabric. Indeed, the ability of these synthetic beings to mirror human behavior challenges understanding of authenticity, blurring the lines between the sentient and the simulacra. As humans venture into this exploration, it becomes clear that while robots may lack true consciousness, their presence and actions have tangible effects on human perception, compelling people to reevaluate the essence of social interactions and the boundaries of their own identity.

These robots, designed to simulate human social cues, to participate in dialogues, and to exhibit a semblance of understanding, do not possess consciousness as people comprehend it. Yet, their gaze still impacts the perceiver, invoking a confrontation with an other that is fundamentally alien. The human subject finds itself mirrored in a visage devoid of traditional life, an uncanny reflection that both alienates and familiarizes. The robot emerges as both a mirror and a rupture, reflecting the human condition while simultaneously deconstructing ontological certainties. This era of human-robot interaction is not just about automation; it is about who humans become in response to it.

The gaze of the other is instrumental in an individual's realization of themselves as a being perceived by another consciousness. With social robots, this gaze is simulated yet impactful, creating a new form of mutual recognition between human and AI and machine. The existence of social robots challenges people to reconsider the nature of the gaze itself—its origin in consciousness or its mimicry—and what this means for the understanding of self and society. What does it mean to exist in a world where the gaze is no longer solely human? What transpires when this transformative gaze emanates not from an organic consciousness but from the inorganic eyes of a social robot?

Moreover, the introduction of social robots into society reshapes the phenomenological structure of the human world. Social robots are not mere tools but active participants in the ontological drama of existence, contributing to the continual redefinition of social roles and relationships. Each robot, with its social veneers and simulated interest, reflects social values, biases, and vulnerabilities. They do not just serve or assist; they participate, and by participating, they alter the fabric of social reality.

As people become increasingly immersed into this era of an organic-synthetic society, the questions are not only how social realities are constructed but also who, or what, participates in their construction. In a world where technology is the new frontier of social stratification, robots are not just machines—they are the newest actors in the drama of inequality. As automation spreads, these mechanical beings become symbols of both hope and displacement, reshaping labor markets, redefining productivity, and reconfiguring the very notion of what it means to be human. The social role of robots is not a question of if they will impact peoples' lives but how—and for whom. Will they become tools of liberation, easing burdens and expanding access, or will they deepen existing divides, serving the privileged while marginalizing the vulnerable? This narrative is about more than just technology; it is about power, access, and the ongoing struggle for equity in a rapidly changing world.

As such, the presence of social robots among people is not merely a technological innovation; it is a philosophical provocation. It beckons people to question, to delve deeper into the nature of these new social interactions. The entrance of robots into the social fabric is not merely an addition but a cultural transformation as society navigates what it means to construct reality when part of this reality is no longer exclusively human.

This book explores these intersections of technology and human experience through the lens of everyday occurrences and will delve into the stories of those who find their reality subtly altered by their interactions with robots. This is more than recording an evolution; it is presenting the emergence of a redefinition of the human narrative in the age of advanced robotics.

ONGOING RE-/DEFINITION OF SOCIAL ROBOTS

Because of the unknowns in human-robot relationships, the world is very much at a nascent stage of exploring the initial introduction of social robots and is also very much co-constructing nuanced cultural realities for how to regard social robots. Additionally, to this date, very little research has been done about the creative influences on the first waves of social robot makers and the factors informing peoples' expectations of their interactions with these first robots in their everyday lives.

While individuals may attribute social qualities to any robot based on their personal interactions and the narratives they construct, this is complemented by the deliberate efforts of engineers and researchers to imbue robots with the ability to understand and reciprocate human social cues. Thus, the distinction between perceived and intended socialness in robots narrows, as the advancements in technology enable robots not just to act but to react in ways that are increasingly sensitive to the nuances of human behavior.

The field of human-robot interaction (HRI) encompasses a range of technologies and methodologies that enhance the way humans and robots interact, fostering trust, comfort, and cooperation. Technological advances in artificial intelligence (AI), natural language processing (NLP), and computer vision combined have made it possible to create robots that respond appropriately to human emotions, gestures, and speech (Breazeal, 2003), opening a possibility for intentionally designed socialness in robots. These technological and research breakthroughs have paved the way for the development of early socially capable robots (Softbank, 2014; "*Jibo, the First Social Robot for the Home*," 2017).

Formal definitions of *social robots* often include factors such as it being a machine or embodied AI that is explicitly designed to engage in acceptable social interactions when situated with humans, and thus exhibits behaviors and functionalities that promote communication, emotional connection, and social engagement (Breazeal, 2003; Dautenhahn, 2007; Duffy et al., 2000; Fong et al., 2003). Additionally, a social robot definition often means that while they may receive instructions or commands from people, the robots also have a degree of autonomy and intelligence that allows them to independently make decisions, interact with the world, and navigate social situations (Fong et al., 2003). According to Fong et al. (2003), *socially interactive* robots exhibit the following humanlike characteristics: express and/or perceive emotions, communicate with high-level dialogue, learn models of or recognize other agents, establish and/or maintain social relationships, use natural cues (gaze, gestures, etc.), exhibit distinctive personality and character, and may learn and/or develop social competencies.

So, *socialness* in robots has traditionally referred to the extent to which a robot exhibits traits, behaviors, or characteristics that are perceived as social by humans, including the ability to engage in interactions that are typically associated with human social behavior. This capability can encompass verbal and non-verbal communication, recognition and expression of emotions, establishment of social rapport, and adherence to social norms and etiquette. The concept is grounded in the field of social robotics, which intends to build robots capable of functioning in a human social environment in a manner that feels natural and comfortable to people (Breazeal, 2004). Social robots are typically being defined by these design factors that are meant to be engaging and even friendly, often with anthropomorphic or zoomorphic features, to facilitate natural human-robot interaction (Dautenhahn, 2007). There is a focus on the design of social robots involving the integration of social cues and affordances that communicate the robot's *readiness* to engage in social interaction, its attentiveness to social dynamics, and its ability to adapt to the social context of the interaction (Bartneck & Forlizzi, 2004).

Yet, there is a growing base of research that demonstrates people may also define a robot as *social* when they project socialness onto a robot not designed for this purpose. Therefore, via the *human gaze*, I expand the definition of *social robot* to include any robot that people interact with socially regardless of the design intent. This expanded definition acknowledges that the person interacting with the robot that they regard as *social* centers on that person's projected narrative onto the robot. This perception may include the misconception that the robot has capabilities that it does not have (Carpenter, 2016; Zawieska et al., 2012).

Recognizing the multifaceted nature of social robots, the developers must consider the implications of their evolving role within society. While the inherent socialness of robots may initially be a projection of human expectations and narratives, this perceived socialness is beginning to merge with the intentional design of robots as companions and helpers. This convergence of perceived and designed social abilities is particularly pertinent as to consider with the burgeoning application of social robots in critical sectors. By integrating robots that are both designed to be social and those perceived as such by humans, societies will begin to see a landscape where robots are not only tools but partners—capable of meeting the complex emotional and practical needs of various populations, including the elderly. The expanding capabilities of these robots are poised to deliver profound benefits, exemplifying the tangible impact of social robots in addressing real-world challenges and human needs.

Thus, the first waves of social robots are currently demonstrating their potential for diverse applications, including healthcare, education, therapy, and entertainment (Broadbent, 2017). For example, the aging population in many countries has created a demand for assistive technologies that can provide companionship and support to older adults (Heerink et al., 2010). It is anticipated that this increasing number of older people will need nursing and support for their general needs by upgrading their quality of life and that social robots have the potential to address social isolation and aid to older individuals.

STORYTELLING AND EXPECTATIONS OF A ROBOT

There is no universal agreement about how to define the word *robot*, yet the word has become an integral part of contemporary discourse due to the advancements in robotics and artificial intelligence. However, the word's meaning is nebulous, as it is deeply influenced by cultural contexts, and is therefore often (mistakenly) used as if the word is a settled matter, as if everyone involved in the discourse has a common understanding of what a robot is conceptually.

In the first-person interviews here, people clearly have many different ideas of what a robot *is*, *could be*, or *should be* in relation to themselves. For the most part, the experiences people share here revolve around humanlike robots, or robots that have design cues built in to imply a humanlike appearance and respond or behave in ways that suggest humanlike intelligence.

In this book, I will use the term *robot* to refer to a machine that physically embodies artificial intelligence and that interacts with and learns from its environment. I mean this definition to be purposefully flexible and as a basis of understanding what *robot* means, unless otherwise expanded upon in its abilities or limitations.

However, the word *robot*, or *robota*, was first used in Karel Čapek's 1921 play *R.U.R.* (2004). The word roughly translates to *forced labor* or *drudgery* and was used to describe artificial, humanlike beings (Philmus, 2001) that were organic, but assembled rather than born. The play presents a dystopian future where robots—initially designed to serve humanity—eventually rebel against their creators, leading to the extinction of born humans. In addition to being a contemporary political criticism and reflection of the political turmoil in which Čapek lived in a post-WWI Europe, the play also highlights deeply rooted cultural anxieties surrounding the continued industrialization of what had previously been human (or animal) labor. Thus, the term *robot* has often been associated with mechanized labor and the potential displacement of human workers (Turney, 2022).

The concept of *robots* has evolved significantly since its inception in Čapek's play. The idea is continually shaped by historical and technological developments, as well as the cultural context in which these changes occur. It was in the latter half of the 20th century that the concept began to encompass a broader range of technologies, from *drones* (Drones & Robots, 2024) to *droids* (Lucas, 1977).

This evolution of the concept of what a robot *is* demonstrates its conceptual malleability and the influence of cultural factors on its meaning. The term has been used to express both hopes and fears regarding the implications of technological progress, reflecting the shifting cultural attitudes and values that shape its interpretation. Accordingly, the word *robot* is also culturally influenced by its representation in various forms of media, such as literature, film, and television. These cultural products often present robots as both a source of fascination and a potential threat, further illustrating the complex nature of the term and its meaning.

For example, science fiction literature has played a significant role in further shaping the concept of robot with authors like Isaac Asimov and Philip K. Dick exploring themes of artificial intelligence, autonomy, and human-robot relationships. Similarly, Hollywood films like *Metropolis* (Lang, 1927), *Star Wars* (Lucas, 1977), *Blade Runner* (Scott, 1982), and *Ex Machina* (Garland, 2014) have contributed to cultural beliefs about robots by presenting portrayals of these artificial beings that explore how people will incorporate social robots into their lives. These robot narratives influence mental models that people have developed in large part to stories like these, their first introduction to the idea of robots in their everyday lives. These cultural representations not only influence how the term *robot* is understood but also shape the expectations and attitudes surrounding the development and integration of robotic technologies into society. Therefore, the very concept of what a *robot* means is far from a universally agreed-upon term and instead deeply embedded in and influenced by cultural contexts, evolving alongside historical, social, and technological developments.

While the cultural landscapes are ever-changing, the representation and conceptualization of robots have undoubtedly been influenced by this dynamism. As societies continue to be shaped by the narratives presented in literature, film, popular media, video games, and advertising, the collective understanding of robots evolves, reflecting the diverse ways in which these artificial entities can be envisioned and realized. It is within this framework of cultural flux that the idea of a robot becomes a tapestry, woven with threads of past lore and present-day storytelling, each addition altering the pattern of the whole.

In recognizing that the concept of a robot is an ever-shifting construct, shaped by the cultural milieu from which it arises, people are better equipped to understand the nuances of how these technological beings fit into the human narrative. It is here, at the intersection of culture and innovation, that the true essence of what a robot signifies is continuously forged and redefined.

The concept of something being culturally *situated* (Suchman, 1987) refers to the idea that meanings, values, and practices are embedded within specific cultural contexts. In other words, a culturally situated concept is one that is influenced by historical events, social values, traditions, media representations, and other factors unique to a particular culture or group of people. Thus, the meaning of a culturally situated term like *robot* is fluid and can vary across different contexts. Acknowledging the culturally situated nature of robots is essential for a more accurate inter-subjective understanding of the complex interplay between individuals, societies, and the ideas they share. The meaning of a robot is not fixed but shaped by cultural factors, an interplay of the diversity of human experiences and perspectives.

Understandably, in a modern world where people consume content in the form of everything from art to advertising, media representations of robots play a significant role in shaping the culturally situated concept of what associations there are with the idea of robots. The beauty and terror of storytelling is it shows the world through the lens of someone else's ideas or experiences. American literature, film, and television have often presented robots as both a source of fascination and a potential threat, influencing public perception and understanding of the term. Science fiction writing, such as Isaac Asimov's works, has explored themes of artificial intelligence, autonomy, and human-robot relationships, while films like *Metropolis* (1927), *Artificial Intelligence: AI* (Spielberg, 2001), and *After Yang* (Park et al., 2022) have presented portrayals of robots that reflect widely differing cultural attitudes and values.

Comic books, manga, and anime also often serve as philosophical mirrors to technological advancement. Throughout the years, in these mediums, robots have evolved from mere mechanical entities to complex characters that reflect society's values, anxieties, and aspirations and influences our perception of technology, humanity, and the future. As robots have become more integrated into our lives, these illustrated mediums have depicted this progression. For instance, in Japan during the 1950s, Osamu Tezuka's *Astro Boy* introduced a friendly, childlike robot that embodied themes of hope and excitement about the potential of technology that reflected a post-World War II era interest in science and progress (Fernandes, 2009; Mičulková, 2016; Tezuka Osamu Official, 2024).

Robot representation in these mediums often reflect society's anticipation and fears of the future. American comics like *Iron Man* (Lee et al., 1963; Hogan, 2009) and *The Transformers* emphasize individualism and heroism (Mantio, 1984; Underwood, 2013), narratives that serve as explorations of the potential consequences of unchecked technological progress. Japanese manga and anime like *Neon Genesis Evangelion* (Anno et al., 1995), *Battle Angel Alita* (Kishiro, 1990–1995), and Shirow's cyborgs in *Ghost in the Shell* (1989–1991) explore themes of identity, cybernetics, and existentialism, robot representation that in these mediums frequently raise complex ethical questions about the boundaries between humans and machine and explore the consequences of playing with the power to create life or *lifelikeness*, addressing moral dilemmas surrounding stewardship of the development of advanced technology.

Via film, science fiction, comic books, manga, anime, advertising, and video games, the representation of robots is more than just a technological fascination; it reflects embedded cultural values, ethical dilemmas, moral philosophies, and aspirations. These representations have evolved alongside society, providing a window into human relationships with technology in different situations and visions of the future. As people continue to grapple with the complexities of robotics and artificial intelligence in the real world, these mediums serve as crucial sources of cultural introspection and artifacts of exploration, reminding everyone of the enduring cultural significance of robot representation in popular culture.

These stories shared about robots inform peoples' expectations of their interactions with robots in relation to themselves. Indeed, fictional humanlike robot representations may inspire a sense of humanlike empathy that is projected onto them into the real world, mistaking the robot's narrative or *journey* for something like a humanlike life experience (Ebert, 2005; Barbir, 2021; Carpenter, 2016). Yet, a social robot's *journey* begins and ends with human perception of the robot as a social being; the *journey* always resides in the people interacting with a robot in a way that acknowledges it as a social other.

Recognizing the culturally situated nature of the concept of *social robots* has important implications for the development and integration— or rejection—of robots in society. Acknowledging the ongoing influence of cultural factors on the meaning of the term *robot* and its associations means a better understanding of diverse perspectives and expectations surrounding robotic technologies.

ETHICAL CONCERNS ABOUT SOCIAL ROBOTS ALREADY IN USE

According to the World Economic Forum, Japan's population is the most rapidly aging, with individuals aged 65 years and older already making up almost a third of its population as of 2023; this percentage is the highest in the world and is expected to continue increasing. This demographic shift is a direct outcome of significant advancements in healthcare and economic development within Japan, and it mirrors a growing trend on a global scale (idem.). Consequently, Japan has served as a pioneer in the emerging era of robot care technologies designed specifically for aging populations worldwide, and its experiences will offer valuable insights for other countries that face similar challenges in the future. The aging population in many countries, notably Italy and Finland, has also created a demand for assistive technologies that can provide care, companionship, and support to older adults ("More than 1 in 10 people in Japan aged 80 or over," 2023; Heerink et al., 2010). Yet, Japan's hopeful investment in robots to assist the elderly has not necessarily produced the anticipated, optimistic results yet ("More than 1 in 10 people in Japan are aged 80 or over," 2023; Wright, 2023). Some of the initial challenges identified with Japan adopting widespread use of social robots in care scenarios have included costs of acquisition, leasing, and maintenance, and the limitations of robot capabilities in terms of utility, intelligence, or socialness (idem.).

For instance, robots like Paro, a therapeutic harp seal-like robot, have shown promise in improving the well-being of individuals in healthcare settings when used to provide emotional support and reduce stress and anxiety among patients (Kidd et al., 2006; Shibata et al., 2021). However, when there is an opportunity for emotional attachment to an object like a social robot, that also will mean, in part, there is a prospect for anxiety and distress when someone is separated from that object (Bowlby, 1979; Carpenter, 2016; Wright, 2023). Keeping living, real pets involves factors such as animal caregiving, allergy, disease, biting and scratching troubles, all of which make it difficult to introduce animal therapy in the medical care and welfare facilities. Paro is already in use or testing in nursing homes, veteran hospitals, and other care facilities as a version of animal therapy for patients with a range of emotional and physical challenges, including dementia and post-traumatic stress disorder (PTSD) (Jøranson et al., 2016). Furthermore, NAIST plans to expand Paro's use-case scenarios to children with developmental problems (2014). Thus, we see a robot that is designed specifically to fulfill emotional needs in human interactions and targeted at vulnerable populations, such as children and senior citizens.

Social robots may indeed someday have the potential to address social isolation and aid individuals who want or need care, but critics argue that increased reliance on social robots may lead to a reduction in meaningful human interactions, and overdependence on technology for companionship hinders the development of genuine relationships (Turkle, 2020). While social robots have the potential to offer benefits, it is important to acknowledge and address the potential challenges that have been proposed or predicted.

There are arguments that prolonged interaction with social robots will create overreliance and increase social isolation and that a reliance on social robots for companionship or social interaction may lead to a reduction in human-human interactions (idem.), and thus become isolated and miss out on benefits of genuine human connections. This line of thinking is frequently premised on the idea of robots substituting or limiting meaningful human relationships instead of the notion of robots complementing an individual's social or care needs.

The deployment of social robots raises other ethical concerns, particularly regarding user privacy, data security, and emotional manipulation (Darling, 2021; Sharkey & Sharkey, 2020). Although social robots can exhibit empathy-like behaviors, their understanding of human emotions is limited compared to humans and they do not have anything resembling humanlike empathy. Furthermore, in the course of their interactions with users and their environments, social robots collect and store personal information considered sensitive, potentially compromising user privacy (Darling, 2021). Additionally, there is a risk that social robots can be manipulated or misused to deceive or exploit vulnerable individuals. Imagine the situation of robots produced by global entities that encourage users to covet or but certain products, follow particular ideologies, or otherwise employed for dark uses.

In some cases—perhaps especially with the robots aimed at vulnerable populations—critics warn that the introduction of social robots into human social spaces may even lead to a devaluation of human-human interactions (Sweeney, 2023; Sharkey & Sharkey, 2020; Turkle, 2020). They believe there is a risk of perceiving humans as interchangeable with robots or reducing social interactions to transactional exchanges, which eventually may undermine the essential aspects of human relationships, such as empathy, understanding, and emotional connection.

Then, consider the emotional attachments that people potentially form with robots can have both positive and negative consequences. If robots are designed to replicate socialness or companionship interdependency, then the absence of that robot may create anxiety or distress for its owner, as any emotional loss might. While these robots have the potential to enrich lives by providing companionship, support, and assistance in various contexts, such as eldercare, the *human gaze* raises an opportunity for reflection about the nature of cultural relationships with robots, the potential for manipulation or exploitation of social robots (intentional or otherwise), and the responsibilities of roboticists and policymakers when shaping these interactions.

EXPLORING EMERGING CULTURAL BELIEFS ABOUT SOCIAL ROBOTS THROUGH PERSONAL NARRATIVES: A PHENOMENOLOGICAL AND SOCIAL CONSTRUCTIVIST APPROACH

Understanding emerging cultural beliefs is crucial, as it provides insights into the dynamic nature of societies and the factors that shape human experiences. As such, case studies of personal narratives offer a valuable means of exploring these beliefs, as they allow for an in-depth examination of individual experiences within specific cultural contexts. There is great value in using case studies of personal narratives to explore and describe emerging cultural beliefs from a phenomenological framework, when employed with methods rooted in ethnography to achieve a nuanced understanding of the evolving cultural landscape.

Phenomenology is a philosophical and methodological approach that focuses on the study of human experiences and the meanings individuals ascribe to them. This conceptual framework is particularly well-suited for exploring personal narratives, as it emphasizes the importance of understanding individuals' subjective experiences within the context of their specific cultural environments. Adopting a phenomenological approach allows insight into the complex ways in which individuals interpret, negotiate, and embody emerging cultural beliefs (Husserl & Moran, 2012; Reeder, 2010; Cilesiz, 2011).

As a philosophical approach, *phenomenology* centers on the study of consciousness and the structures of experience as they present themselves to an individual. It seeks to understand and describe the essential structures of human experience by focusing on the first-person perspective (Dewey, 1938; Cilesiz, 2011; Coeckelbergh, 2011a; Husserl & Moran, 2012), and not simply behaviors. Thus, this approach includes both the *subjective* and *intersubjective* dimensions of experience, which allows for a rich and nuanced understanding of the world as it is lived. By examining the subjective dimensions of experience, phenomenology emphasizes the importance of the individual's perspective, the unique ways in which they experience the world, and the meanings they ascribe to their experiences. Key concepts in phenomenological perspective include *intentionality*, the *lived body*, the *lifeworld*, and the *epoché* or *bracketing* (Husserl & Moran, 2012; Embree & Nenon, 2012; Gallagher & Zahavi, 2023).

This foundational concept of the *lifeworld* refers to the world of lived experience inhabited by conscious beings, encompassing the way phenomena, including events, objects, and emotions, manifest themselves in our conscious awareness within our everyday existence. It encompasses the entirety of individual subjective experiences, from the most mundane to the profound, and is the stage upon which phenomena come into view. In the *lifeworld*, attention is naturally directed toward the content of experiences—what is perceived, felt, and encountered—without reflecting on the processes or mechanisms of perception itself. In other words, Husserl's focus on the lifeworld underscores the intrinsic relationship between conscious beings and their immediate environment (Husserl & Moran, 2012), or situatedness. Phenomenology further conceptualizes the lifeworld as *pre-reflective*, focusing on what people perceive rather than how they perceive it (Husserl & Moran, 2012; Merleau-Ponty, 1962).[1]

Furthermore, within the realm of phenomenology, the notion of *evidence* takes on a distinctive character. "In phenomenological thought, evidence is not merely a signifier of objective proof; rather, it *represents the subjective achievement of truth*" (Merleau-Ponty, 1962, p. 168). This conceptual framework puts primacy on delving into the intricate relationship between individual perception, intentionality, and knowledge. In this context, the concept of *evidence* takes on a unique role, diverging from the conventional understanding of evidence as a mere indicator of factual truth. Thus, in the phenomenological framework, the term *evidence* encompasses the idea that truth is not an abstract, external entity waiting to be discovered. Instead, it is an active and dynamic process within consciousness. *Evidence*, in this sense, is not a passive observation of facts but a participatory engagement with the world. It entails the successful presentation of an intelligible object, a presentation in which the truth of that object becomes manifest through the act of evidencing itself.

To grasp the essence of the phenomenological notion of evidence, one must also consider the role of *intentionality*—the inherent directedness of consciousness toward objects. In phenomenology, evidence is intricately linked to the presentation of an object within the realm of intuition. The act of evidencing involves not merely the sensory perception of an object but also its apprehension as intelligible, and it is in this intelligibility that truth comes to the fore.

Moreover, a phenomenological lens asserts the profound recognition of *the body*'s centrality in human consciousness and lived experience. It asserts that the body is not a mere physical entity but the very medium through which people perceive, act, and engage with the world (Merleau-Ponty, 1962). This framework highlights an inseparable connection between the body and selfhood. The body's role extends beyond individual perception to intracorporeal engagement with others, fostering an ethical responsibility and an enriched understanding of human existence. For instance, the way we move and interact with the world contributes to our sense of self. Our body's gestures, posture, and actions are not mere physical behaviors but expressions of our subjectivity. Phenomenology thus illuminates the intricate relationship between the body and the self, emphasizing that the body is not just a vessel but a constitutive element of our identity.

Social constructivism posits that knowledge and reality are constructed through social interaction, culture, and language (1978, Vygotsky; Vygotsky, 1986). Thus, this perspective emphasizes the role of social, historical, and cultural contexts in shaping human experience and knowledge. Central to the *social constructionist* framework is the idea that *meaning* and *understanding* are co-constructed through the dynamic process of social interaction which in turn influences the ways individuals make sense of the world in a process called *intersubjectivity*, or the social co-construction of reality, and the important role of language and culture in shaping human experience (Berger & Luckman, 1991). By foregrounding the relational aspects of human experience, social constructivism and constructionism offer a valuable lens through which to examine the ways in which social processes and contexts shape individual and collective understanding. This perspective posits that cultural knowledge and beliefs about robots are evolving constructs that are constantly reinforced or revised by their expectations based on media (e.g., fictional narratives), as well as the influence of real experiences with robots and the mutual gazes of other people using robots.

Phenomenology and social constructivism share complementary qualities that can be harnessed for interdisciplinary research. Both approaches emphasize the role of human experience, meaning-making, and the dynamic interplay between individuals and their environments in shaping knowledge and reality. Social constructivism argues that through processes of *socialization, externalization,* and *objectivation,* individuals collectively construct their social worlds, embedding shared meanings and values into the fabric of society (Berger & Luckman, 1991; Gergen & Gergen, 2004). In this way, individuals collectively create and sustain their social world through ongoing processes of interaction, communication, and mutual agreements, so social reality is a dynamic and fluid entity shaped by the meanings and interpretations attributed to it by society's members.

Also central to the thesis of social constructivism is the role of *socialization* in shaping individuals' perceptions of reality, or the process through which individuals acquire the knowledge, beliefs, and norms that enable them to participate in society (Berger & Luckman, 1991). This process occurs through various institutions, including family, education, religion, and media. This *socialization* instills individuals with the shared meanings and values that underpin the social world, allowing them to navigate and make sense of their environment.

Via socialization, individuals learn to identify with specific social groups based on shared cultural characteristics, such as ethnicity, religion, gender, and nationality (Lorber & Farrell, 1991; Lopez, 1991; Gergen, 1995; Gergen & Gergen, 2004; Lorber, 2021). These social identities influence individuals' self-concept, values, and behaviors, contributing to the complexity of their social reality. The concept of socialization's role in social constructivism is evident in numerous real-world examples. For instance, gendered robot roles and stereotypes are constructed through socialization processes that dictate expected behaviors and characteristics (Butler, 1999; Carpenter et al., 2009; Lorber, 2021; Nomura, 2017; Siegel et al., 2009). Similarly, cultural norms and values regarding family structures, religious beliefs, and political ideologies are transmitted and upheld through socialization (Kagitcibasi, 2017).

Culture provides the tools and meanings that shape human cognition, and individuals actively engage with cultural artifacts to construct knowledge (Vygotsky, 1978). Within the framework of social constructivism, the determination of cultural groups involves recognizing the role of shared meanings, practices, and cultural tools in shaping identity and belonging.

Cultural groups are not defined solely by objective criteria but are constructed through shared experiences and interactions.

Thus, cultural groups can be described by these shared practices, traditions, rituals, and customs that members engage in collectively. These shared activities contribute to a sense of belonging and identity within a cultural group (Bruner, 1990). Shared language patterns, dialects, and communication styles also help individuals identify with a particular cultural group (Gee, 1999; Jörgensen, 2016; Nass & Brave, 2005). Membership in this group shapes individuals' self-concept and worldview, influencing how they perceive themselves and others (Wertsch, 1991).

The social constructivist framework also introduces the key concepts of *externalization* and *objectivation* to elucidate the process by which social reality is constructed. *Externalization* refers to the human capacity to create and externalize meaning through language, symbols, and actions. In other words, individuals express their thoughts and intentions through communication and interaction with others. *Objectivation*, on the other hand, involves the transformation of externalized meanings into objective, tangible structures within society. This process solidifies shared meanings into institutionalized practices, norms, and institutions, reinforcing the stability of social reality. Over time, individuals often take their social reality for granted as the routines and practices of daily life become so ingrained that people rarely question or challenge the underlying assumptions that shape their worldviews. This taken-for-granted nature of reality not only maintains social stability but can also limit individuals' ability to recognize the constructed nature of their reality, and I argue this is one way the idea of the *human gaze* offers an exercise in self-reflexive theory and praxis.

Phenomenology and social constructivism also adopt an epistemology that posits that knowledge is actively constructed by individuals and is subjective in nature. This shared orientation highlights the importance of human experience and meaning-making in understanding the world, allowing for rich explorations of the complex interplay between individual subjectivity and social context (Bruner, 1990).

Together, the approaches of phenomenology and social constructivism lenses can be fruitfully combined to create a rich, in-depth understanding of complex phenomena. Phenomenological methods, such as *epoché*, reflective practice, and descriptive analysis, can be used to explore the lived experience of individuals, while social constructivist techniques can be employed to examine the social and cultural contexts that shape those experiences.

INTRODUCTION TO THE HUMAN GAZE

Humans learn, in part, by observing how others behave as way to shape their own initial mental models (Bransford et al., 1999; Vygotsky, 1978, 1986; Derry, 2013). As humans, people can imagine a life in a way that they aspire it to be in the future. How people determine those subjective and very personal aspirations are influenced in great part because they covet what they see modeled, internalize it with cultural expectations, and then imagine new mental models of what they hope for themselves. To that end, to reach those goals, they choose and shape tools to move toward those aspirations. Thus, this combination of human imagination, coveting, projection, survival, and curiosity drive people to develop tools based on what they think they know and what they expect will be the outcomes. However, this ability to forward-think and imagine next generations of tools is also based on the successes, failures, utility, or joy the previous tools brought them. Therefore, in turn, the tools and artifacts shape the people that made them.

In this way, tools are not just a reflection of a culture's current state but also a map to its desired future. As cultures navigate through this progression of integrating robots into everyday life, note that as with other technologies, the relationship between humans and the tools they create is reciprocal; while shaping tools with the intent of fulfilling aspirations, these very tools, in turn, shape expectations and expand the human imagination. Entering the realm of artificial intelligence and robotics, this dynamic takes on new dimensions. As makers imprint humanlike cognitive models onto these new machines, robots are not just a type of mirror to reflect an existential version of what it is to be human but it also forges a lens through which people can envision and shape the future.

The *human gaze* is inspired, in part, by Sartre's concept of *le regard*, or *the look*, which plays a central role in his existentialist account of human experience and consciousness (2015). However, the *human gaze* refers to the development of AI—and robots—from a human-bounded set of cultural expectations and human-created data, which is the phase of human-robot interaction we are now. Our initial expectations and perceptions of its output are ultimately based on biases toward seeing AI as interpretations of human culture, understanding, meaning, and desired goals. Therefore, the *human gaze* refers to positioning of AI to humans and using humanlike paradigms of understanding the world as AI R&D models, specifically robots.

Is a human gaze *wrong* in some way? To that I say, no, but there is value in understanding it and the awareness of it, to try and to attempt to leverage its advantages and mitigate its biases when possible or useful. Yes, people should be designing technology to serve their needs, but they should not limit their vision of possibilities of human-robot interactions based on the implicitness the gaze insinuates itself into their being.

Furthermore, I stress that this gaze is not a static state or permanent condition but a philosophical underpinning that changes over time as humans and robots continue to coexist and their social interplay adjusts and transforms.

Consequently, I am interested in how the human gaze applies to AI that is presented as a humanlike social robot, or robots that people *project socialness onto*, and the interaction phenomenon surrounding it.

A summary of the *human gaze* qualities regarding social robots:

1. Humans are the bearer of the gaze by default.

2. Assumes rights and cultural privileges over the robot, and society agrees with this assumption.

3. Value of a robot is determined by its appeal and use to humans, thus reducing robots to objects for human utility.

Furthermore, the gaze may have additional qualities from the point of view of the robot makers:

- Consistently reflects and reinforces society's structures via robot design. By designing social robots in ways that mimic human social constructs, robots help maintain a sociocultural status quo.

And some human gaze experiential qualities or conditions of the person(s) interacting with robots may include the following:

- A perception that the robot is a potential social actor or other.

- Belief the robot other has an inner *vitality* of its own, whether that is technical (AI), organic (innate), or both. This *vitality* is defined as the power or ability to continue in existence, live, learn, or grow.

- The human social actor is invited to identify with the social robot, which becomes an entity that possesses unique experiences, perspectives, and identities. This identification is constructed through the user's perspective and the narrative's unfolding, which privileges the user's viewpoint.

- The relationship of the human self to the robot as other unfolds over time, with interactions and perceptions evolving the *human gaze* dynamically.

A *social actor* refers to an individual or entity that engages in social actions within a society, influencing and being influenced by social structures and interactions. The concept of a social actor is central to understanding how individuals and groups navigate and affect their social environments. It recognizes that these actors have agency, meaning they can make choices and take actions that impact both their own lives and the broader social context. Social actors are not merely passive recipients of social forces but are active participants who interpret, negotiate, and sometimes challenge social norms and structures (Giddens, 1984). Typical social actors include people, groups, organizations, and institutions that participate in social processes and contribute to the shaping of social life.

A *social other* here refers to an individual or group that is perceived as different from and outside of the dominant social group. This concept is central to discussions of identity, power, and social dynamics, as it highlights the ways in which societal norms and structures create and reinforce distinctions between *us* and *them*. The social other is often marginalized or stigmatized, which can result in exclusion, discrimination, or other forms of social disadvantage. The construction of the social other is a process that involves defining what is considered "normal" and "acceptable" within a society, and subsequently identifying those who do not fit these criteria. This process can be influenced by various factors, including race, ethnicity, gender, sexual orientation, religion, and socioeconomic status (Hall, 1997).

In the psychological and philosophical discourse, the *ego* is often understood as the aspect of the self that mediates between the conscious and unconscious and is responsible for reality testing and a sense of personal identity.[2] Sartre, on the other hand, approached the concept of the ego from a phenomenological and existential perspective, particularly in relation to his notion of *le regard* (or *the gaze*). For Sartre, the ego is not an innate repository of the self but is instead constituted externally, acquired through the gaze of the other (Sartre, 2015). This is vividly captured in his analysis of *le regard*, where it is the presence of another person that causes an individual to become suddenly self-aware, reflexive, experiencing themselves as an object in the other's world. This moment of objectification by the other's gaze is pivotal for the development of the self, as it marks the transition from being a purely subjective consciousness to recognizing oneself as an object among other objects within the world.

Sartre's concept of *le regard* highlights the existential angst and loss of autonomy that accompanies one's realization of being an object for other's gaze. The connection between Sartre's *le regard* and the *ego* lies in the recognition that the self's understanding and construction are deeply influenced by social interactions and intersubjectivity of the gaze of the other. The ego is not merely a personal identity crafted in isolation but is dynamically shaped by the ways in which individuals perceive and are perceived by others in their social world.

Thus, referencing the basis of *le regard*, the *human gaze* refers to the positioning of the humans' sense of the self in relation to social robots and the realization of the self through a new ontological gaze. I encourage us to explore AI and social robot research and development methods that shift prior assumptions, thereby offering varied perspectives and stories to explore.

THE WORK'S PROPOSITIONS

Based on the idea of the *human gaze*, the work in this book was guided by the following research propositions:

P1: Robots are still relatively new technologies being integrated into peoples' everyday lives in different roles and contexts of use (i.e., industry, military, or social).

P2: People have expectations about what their interactions with robots will be like based on *stories* (i.e., science fiction, religion, or news stories). These expectations inform how they design, make, represent, and interact with robots.

P3: In turn, the robots created are based in deeply culturally embedded beliefs that relay aspects of the stories makers learned, often by manifesting properties of those stories in robots that perpetuate or reinforce characteristics of those values, beliefs, and expectations.

This work's propositions require a specific focus on comprehensively documenting and describing interactions with and expectations of social robots. To achieve this, I adopted a collective case study approach, which involves simultaneously examining multiple cases to gain a broader perspective on complex issues (Stake, 1995).

NORMALIZING ROBOTS BY CREATING A FAMILIAR OTHER

Accordingly, using the *human gaze*, human consciousness is not a static entity, but rather a constantly changing and evolving process. Building on *le regard's* human-human paradigm for a human-robot model of robot development offers a perspective on the relationship between *subjectivity* and *intersubjectivity* and foregrounds the role of awareness of our interpersonal relations in shaping subjective experience of others.

Sartre's approach emphasizes the importance of perception, consciousness, and the dynamic interplay between individuals and their environments in shaping human experience and highlights the central role of perception and consciousness in shaping human experience. By emphasizing the ways in which individuals perceive and make sense of the world, these concepts provide valuable insights into the processes through which meaning is constructed and experience is constituted.

The *human gaze* also has significant implications for our understanding of human relationships and the dynamics of power and control with social robots. By examining the ways in which individuals manage their self-presentation in social interactions and the influence of *the human gaze* on this process, we can gain further insights into the complexities of human behavior and social dynamics.

Sociologist Goffman introduced the concept of *impression management*, arguing that in social interactions, individuals constantly engage in a process of controlling a narrative of how they are perceived by others to create a favorable image of themselves (2002). The dynamic process of impression management involves the use of various verbal and non-verbal cues to convey a particular identity or role. Goffman's work is heavily influenced by the metaphor of the theater, with individuals acting as *performers* who put on different *performances* for different *audiences*. According to Goffman, impression management is essential for maintaining social order and facilitating smooth interactions, as individuals strive to adhere to the social norms and *expectations* associated with their given roles. Similarly, Sartre's concept of *the look* refers to the moment when an individual becomes aware that they are being observed by someone else and subsequently feels objectified and self-conscious. This experience of being seen by another person can have a profound impact on the individual's behavior and self-presentation, as they become acutely aware of their status as an object in the eyes of the other, intersubjectively shaping a sense of the self.

There is a clear connection between Sartre's *look*, Goffman's *impression management*, and how people shape their interactions with robots based on their awareness of being observed. All three concepts emphasize the importance of social perception and the ways in which individuals adapt their behavior and self-presentation in response to the awareness of an other's attention. The experience of the intersubjectivity of a human gaze can serve as a powerful motivator for impression management, as individuals strive to maintain a particular image of themselves in the eyes of others. Understanding the dynamics of *impression management* and *the human gaze* can enrich our analysis of social interactions and the construction of our identity in relation to robots as we examine the ways in which individuals navigate the complexities of social perception and negotiate the boundaries between their private selves and their public personas, and now, learning to navigate how to design and interact with social robots.

This understanding and awareness of the human gaze toward robots is more than self-reflection upon our own role in how we shape culture and culture shapes us, but to also be able to position ourselves to think critically about ways of considering how we grapple with expanding our center of the universe from us as the model of what robots are and should be.

The sociological concept of the other also provides a valuable framework for understanding the ways in which humans perceive and interact with robots (Crang, 1998; Kahn et al., 2004; Coeckelbergh, 2011a; De Beauvoir, 2023; Kim & Kim, 2013) as a social category. By examining the process of *othering* in relation to robotics, we gain additional relevant insight into the potential social and psychological consequences of human-robot interactions, as well as the broader implications of these interactions for societal dynamics and power relations.

This concept of the *other* here refers to the process by which individuals or groups are perceived as being different or separate from the dominant group or culture (De Beauvoir, 2023). Historically, this process of othering often results in the marginalization, stigmatization, and dehumanization of those deemed as the other or Other (Crang, 1998; De Beauvoir, 2023). Othering is rooted in the recognition that human societies are characterized by a constant process of differentiation and categorization, and this process often involves the social construction of boundaries between different groups, with certain individuals or groups being perceived as other based on factors such as race, ethnicity, gender, or class. Thus, the act of othering can serve to maintain and reinforce existing power dynamics, with those in dominant positions using the other to assert their superiority and maintain control over marginalized populations.

Thus, in recent years, the advancement of social robotics and artificial intelligence has raised similar questions about the interactions between humans and machines, particularly in terms of how we perceive and interact with robots as potential social others. For example, the perception of robots as a social other may contribute to the dehumanization of certain groups, as humanlike qualities (e.g., gender and race) are designed into robots, and these machines are perceived as inferior to humans. More specifically, there is a growing body of work addressing how robotic representations of humans, such as gendered robots, can be designed or used in ways that reinforce negative stereotypes and exacerbate existing social divisions and inequalities (MacWilliam, 2023; Richardson, 2016). It is crucial then to critically examine all these issues to ensure the responsible development and deployment of social robots that complement, rather than harm, human interactions.

There is also a growing body of work that proposes a concentrated examination of the social and psychological consequences arising from human-robot interactions, delving into how humans perceive and engage with robots as potential social entities, while also scrutinizing the tendency to attribute human-like attributes to robots (Kahn et al., 2004). According to this research, individuals tend to view robots as belonging to a distinct category, distinguished from both humans and other non-human entities. This classification leads to othering, wherein robots are perceived as inherently distinct and fundamentally lesser than humans.

The discourse surrounding human-robot interaction necessitates a nuanced understanding of how robots, as perceived others, fit into the social landscapes. According to Benesch (1999), droids serve a dual role akin to a societal reflection, rather than an individual one, operating on a broader scale. In his analysis of Philip K. Dick's 1968 story, *Do Androids Dream of Electric Sheep?*, Benesch closely linked the novel to Jacques Lacan's essay on the mirror stage. In this context, Lacan (1953) contends that the development and validation of one's self-concept hinge on the creation of an other through visual representation, commencing with the encounter of a doppelgänger in the mirror. This symbolic interplay between the self and the robotic other not only shapes individual self-concepts but also reflects and potentially reshapes societal norms and values.

THE VALUE OF PERSONAL NARRATIVES AND A COLLECTIVE CASE STUDY

Using a collective case study approach involved consulting various stakeholders engaged in the ecosystem of human interactions with social robots to collect a rich variety of experiences. Ethnographic methods like in-depth interviews and the analysis of cultural artifacts are widely used in the study of human societies and cultures. These methods enable researchers to attempt to immerse themselves in the cultural environments being studied, allowing for a deep and nuanced understanding of the beliefs, values, and practices that shape individuals' experiences. Furthermore, presenting findings as a collective case study is valuable for delving deeply into complex topics or phenomena to explain, describe, or explore emerging cultural events (Luck et al., 2006; Yin, 2017).

The deeper value of these collected cases lies in the transferability of their phenomological findings, and their thick descriptive properties. While human experiences are uniquely situated, by understanding more through identifying and describing how these unique journeys relate to or deviate from the propositions outlined here, there is an opportunity to gain valuable in-depth insights unique to the time this is situated in.

All of the interviews included here were one-to-one and conducted via combinations of Zoom (transcribed), asynchronous cloud-based documents, email, and direct messages. I specifically chose to conduct informal and semi-structured interviews, which enable flexibility during data collection while remaining grounded in a particular framework (Gill et al., 2008). Questions were chosen according to the themes of interest in this study and the individual's area of expertise and experience. In this approach, I used open-ended questions, specifically telling participants they could write without a word limit and without an expectation of word count. Additionally, I frequently included follow-up questions such as "why" or "how" to clarify responses and check their meaning. This approach allowed me to maintain a structured template for the discussion while also remaining flexible to delve into new topics brought up by the participants (Dearnley, 2005; Magaldi & Berler, 2020).

It is also critically important to discuss how I curated the cases (da Mota Pedrosa et al., 2012). The interviewees here are, by their nature, chosen or curated and gathered generally because they are representatives of a phenomenon; they illustrate an example of a larger population, theoretical framework, hypotheses, or propositions. Interviewees were chosen in part because their experience is relevant to the propositions cited here, and are presented as an exploration of emerging cultural shifts, and as such, are viewed as critical cases (Gammelgaard, 2017). The people I requested to interview each had a condition where they are uniquely situated within a prolonged human-robot interaction, which could be any variety of ways, such as through robot design, art, research, or personal relationship. Additionally, the people sharing their stories here made themselves accessible to the process and open to publicly sharing their accounts. Each interviewees' responses are presented in whole, unedited, to present their individual style, tone, and voice as much as possible. Finally, the approach here is an exploration of community members and artifacts from a "bounded system" of interest through detailed, in-depth data collection involving multiple sources of information, data collection, and analysis strategies (Creswell, 2013). The outcome from this collection is a larger case description composed of case-based themes to better understand the human-robot phenomenon and sociocultural issues under study.

However, case studies and interviews are also guided by restrictions inherent to many forms of human-centered research. For example, it is important to note that there are limitations to presenting case studies as an exemplar of phenomenon; rather they should be viewed as unique, subjective, and rich illustrations of human experiences that demonstrate experiential phenomenon. And, admittedly, one limitation of interviewing people is sometimes the inability to access desired participants, which may be due to factors such as language barriers, their distrust of researcher topics, or simply logistical challenges. Collecting first-person narratives through multiple methods, as they were here, include in-depth interviews, written self-reporting, and researcher follow-up fact-checking to ensure that they capture a diverse range of perspectives and experiences.

Furthermore, I feel strongly that my own positionality is transparent. On that note, I specifically use the word *curate* as a form of acknowledging my own role in the collection of interviews here. Triangulating artifacts such as interviews, relevant images (i.e., robots, AI, advertising, and film), and archival primary source material scaffolds the validity I present of the relationships between major themes and the theoretical framework presented. While case studies are limited in their scope and generalizability of findings, it does not detract from the finding's *transferability*, or a valuable way to learn from peoples' own words about their culturally situated experiences. Thus, my decision to include the entire responses from interviewees was to intentionally platform the whole of their unique situations, beliefs, and experiences as they chose to present them.

This holistic perspective is essential for uncovering the multifaceted nature of the nuanced subject under study. Certainly, research questions such as the ones proposed here involve unique or atypical cases that are challenging to study through traditional research methods, and interviews are a way to explore such cases, shedding light on exceptional situations or outliers. This work should be regarded as generative, examining subjective experiences of how such a diverse group of people interacted with and think about robots.

Personal narratives in cases are valuable tools for exploring emerging cultural beliefs, as they provide rich and detailed accounts of individuals' experiences within specific cultural contexts. The in-depth interviews here involve the detailed examination of an individual, group, event, or phenomenon of human-robot social interaction within its real-life context. One of the key benefits of incorporating first-person narratives into ethnographic research includes capturing the individual's subjective perspectives. Interviews also offer a unique window into the contextualization of the individual's thoughts, feelings, and interpretations of their experiences with robots and artificial intelligence, allowing readers to better understand their underlying beliefs and values (Bates, 2004). Furthermore, personal narratives and cases offer a unique window into the complex interplay between individual agency and social structures, highlighting the ways in which individuals both shape and are shaped by their cultural environments.

Case studies also offer an ethical means of studying certain sensitive phenomena, a non-intrusive way to explore and analyze complex human behaviors and social issues. These narratives also provide rich and detailed data, containing extensive information about the specific context in which an interaction took place, as well as the individual's emotional and cognitive responses to the encounter. This information is the invaluable detail in identifying peoples' dynamic beliefs and expectations about human-social robot interaction. As individuals share their personal experiences and perceptions, there is a rich description and identification of emerging cultural beliefs and expectations about human-robot interaction.

NOTES

1 Husserl also distinguishes the isolation of *essences*, the invariant features and structures underlying phenomena, and the precise description of these

essences. Through the isolation of such essences across various experiences, phenomenology aims to uncover the qualities that lend a distinct character to each experiential phenomenon. However, to achieve this objective, it is essential to adopt a specific attitude, one that necessitates the suspension, or *bracketing*, of presuppositions and judgments, allowing for an unobstructed view of the lifeworld to emerge. This critical attitude is known as the *epoché* (Husserl, 1962; Husserl & Moran, 2012). These factors of a phenomological approach to this work may apply to the lens of myself as researcher, the interview subjects, and the readers of this book.

2 In psychoanalytic theory, Freud delineated the *ego* as one of the three components of the psyche, alongside the *id* and the *superego*, playing a crucial role in managing the demands of both the *id* and the external world (Freud, 1923). Freud described the ego as the part of the personality that is seen by others, the *self* that one presents to the world; it develops from the *id* and ensures that the impulses of the *id* can be expressed in a manner acceptable in the real world, and operates according to the reality principle, working to satisfy the id's desires in realistic and socially appropriate ways (Freud, 1923).

CHAPTER 2

Robots in the family

INTRODUCTION

In recent decades, there has been a significant focus of research and development of social robots that aid aging populations as the need increases for effective and affordable solutions to support the growing number of elderly individuals worldwide (Bemelmans et al., 2012; Lazar et al., 2017; McGowan, 2024; "More than 1 in 10 people," 2023; Nakatani, 2019). Special considerations about developing social robots for aging populations include whether people will accept robots as aids, the potential for human emotional attachment and dependency, the role of robots in preserving human dignity and autonomy, the impact on human relationships and social connections, and the potential for unintended consequences and misuse of such technologies (Breazeal et al., 2019).

These robots are not meant to replace human caregivers or the emotional support of adult children but to supplement care, provide additional social opportunities, and offer ways to connect and stay stimulated (Darling, 2022). Yet, they are sometimes presented as things that can be used in the family home (CataliaHealth.com, "*Leading Remote Care Management*," 2024).

Thus, the design of embodied social robots designed to care for older people specifically raises ethical considerations regarding the owner's special vulnerabilities to potential emotional manipulation and decreased autonomy and privacy (Sparrow & Sparrow, 2006; Syrdal et al., 2007). Specifically, there are privacy concerns when robots collect and process personal data during interactions with users who may not have a clear understanding of what they are consenting to (Turkle, 2011).

 DOI: 10.1201/9781003170457-2

For all these reasons, the story that unfolds in the 2013 science fiction film *Robot & Frank* has become an illustrative touchstone for popular human-robot discourse and academic papers on the topics (DeFalco, 2016; Teo, 2020; Ornella, 2015; Koistinen, 2016; Seelman, 2016). The narrative of this movie serves as an especially salient illustration of many cultural factors we are negotiating now with the introduction of the first robots into our most personal and private space, our homes, and determining how or if we integrate robots into our friends and family social circles.

Robot & Frank tells the story of the protagonist, Frank, an older man who is struggling with memory loss and loneliness. In this setting of a not-too-distant future, it is shown that it is commonplace for people to have companion or servant robots following them around like personal assistants, with the robots appearing as variations of humanlike morphology while still clearly machinelike. After some debate among Frank's adult children, they eventually decide to provide him with a robot caretaker, simply referred to as Robot, to help Frank with daily tasks and act as companionship. At first resistant to the robot's presence, Frank eventually forms a friendship with it. Together, Frank and Robot embark on a series of adventures, and as their bond deepens, Frank begins to rediscover the joys of his life as it is in his present, the importance of human connection, and ultimately finding some solace in his newfound friendship with the robot.

The film raises important questions about many of the ethical implications of designing robots for social companionship and assistance to aging and other vulnerable populations, offering a valuable narrative that contributes to the ongoing debate surrounding the use of robots as caregivers and companions for the aging population, children, or anyone with special physical and emotional needs that needs in-home assistance. Therefore, this chapter adds insights into the dialogues regarding robots and aging and vulnerable populations in a new way by exploring the genesis of the story of *Robot and Frank*, itself, a fable that explores—and proposes—the new considerations that social robots may introduce into family dynamics and peoples' everyday lives.

By examining the inspirations of the storytellers, it is possible to gain insights into their expectations of how they believe the audience will interpret the robot's role. Using the idea of the *human gaze* is one way to critically examine how these expectations and models of *robotness* are formed and internalized and then released into the larger consciousness of society via their creative output, here in a real-world robot of ASIMO and the storytelling of *Robot & Frank*. Thus, this chapter includes interviews with Takeshi Koshiishi, the designer of the robot ASIMO; Christopher Ford, *Robot & Frank's* screenwriter; and Rachael Ma, the actor embodying the character of Robot. Their interviews give insights into how they bring their interpretations of *robotness* into a cohesive story, each contributing to the ecosystem of understanding that meaning in very different ways.

Their responses to my interviews, in turn, inspired me to pick these themes for discussion in this chapter: the potential for positive emotional attachment to robots, as well as pitfalls of such attachment; the significance of robot embodiment and movement; and the Uncanny Valley's proposition applied to othering of robots.

EVERYDAY SOCIAL ROBOTS, ATTACHMENT, AND AGING AS A VULNERABLE POPULATION

The global population is aging at an unprecedented rate due to increased life expectancy and declining birth rates. According to the United Nations Department of Economic and Social Affairs (2019), the number of people aged 65 years or over is projected to double to 1.5 billion by 2050. Additionally, there's a swiftly growing disparity between the rising number of elderly individuals requiring care and the diminishing pool of available caregivers (Levinson, 2024). This demographic shift presents significant challenges in healthcare, social services, and support for the elderly. In the United States, the bulk of eldercare work is unpaid and predominantly falls on economically disadvantaged individuals, which in turn exacerbates their financial instability, resulting in an increased likelihood of poverty as they age (Forden et al., 2023). Some research suggests that social robots offer promising solutions to some of these challenges, such as healthcare and physical assistance, as well as cognitive engagement, social support, and companionship opportunities. However, the integration of social robots in the care and support of older adults is not without challenges. Ethical considerations, such as privacy, autonomy, deception, dependency, and the potential reduction in human contact, need to be addressed (Sharkey & Sharkey, 2012).

This sort of aid could be particularly important in rural or underserved areas where access to healthcare providers is limited. Among the proposed tasks for social robots serving older adults is reminding them to take medication, providing information and updates on health conditions, and facilitating telehealth services. An analysis by Vandemeulebroucke et al. (2018) highlighted the potential of robots in supporting elderly care by offering reminders for medication and appointments, thus promoting their independence and adherence to treatment plans. Moreover, for older adults facing mobility issues, robots equipped with physical assistance capabilities can also aid in daily tasks, such as fetching objects, opening doors, or even helping with physical therapy exercises. The development of robots with capabilities to support these physical tasks can potentially enhance the quality of life and autonomy for elderly individuals (Fiorini et al., 2021).

Engaging in cognitive activities is also crucial for maintaining mental agility in older adults, and social robots can facilitate cognitive games, storytelling, and other stimulating activities to help keep the mind active. A study by Wada and Shibata (2007) using Paro (Personal Assistive RObot), a furry, zoomorphic robot designed to mimic a tactile, purring baby harp seal, found that interaction with therapeutic robots could lead to improvements in the overall qualitative aspects of elderly residents in care facilities. There have been various studies emerging in the domain of social robots used to alleviate loneliness or provide companionship in eldercare, and the encouraging findings warrant additional and continued research (Chen et al., 2018; National Institute of Advanced Industrial Science and Technology, 2004; Pu et al., 2018; Shibata, 2012). One of the primary ethical concerns is the autonomy of elderly individuals and their ability to give informed consent regarding the use of social robots. For some elderly users, cognitive decline may impair their capacity to fully understand the implications of interacting with a robot, raising questions about consent (Sharkey & Sharkey, 2012). It is crucial to ensure that elderly individuals or their legal guardians are provided with clear, accessible information about the robot's capabilities and limitations.

Social isolation and loneliness are significant issues negatively affecting the elderly, contributing to various mental health issues, including depression and anxiety. Social robots can entertain and stimulate people by providing companionship, engaging in conversation, and participating in activities to alleviate feelings of loneliness (Salichs et al., 2016). A study by Robinson et al. (2023) demonstrated that social robots could significantly reduce loneliness in adults by providing companionship and enabling social interaction. Nevertheless, the use of social robots to mitigate loneliness among elderly populations raises several ethical considerations that require careful examination. While social robots can provide companionship, there is a risk that their use might lead to increased social isolation by reducing human-human interaction. Others caution that relying on social robots as a substitution for human caregivers or family members can exacerbate feelings of loneliness and neglect among the elderly (Sparrow & Sparrow, 2006).

Unlike Paro, the design of social robots more often incorporates human-like features to facilitate engagement (Phillips et al., 2018). This raises questions about deception and the potential for users to develop emotional attachments to machines that cannot reciprocate feelings in a meaningful way, potentially leading to emotional harm (Turkle, 2011). Furthermore, critics warn there is a risk that reliance on social robots for companionship and care could lead to dependency, which may impact the psychological well-being of elderly users (Sharkey & Sharkey, 2012).

The human tendency to anthropomorphize robots can lead to the formation of bonds and emotional attachments, blurring the social distinctions between humans and machines (Reeves & Nass, 1996; Nass & Moon, 2000; Rosenthal-von der Pütten et al., 2014). The perception of how a robot is embodied or outwardly designed can promote feelings of empathy, trust, and companionship, influencing human attitudes and behaviors toward robots (Eyssel et al., 2012). For example, Cynthia Breazeal's creations of the social robots Kismet and Jibo were intentionally designed with child-like features to naturally elicit caregiving responses from humans. This design encourages humans to interact with the robots as partners, facilitating the development and refinement of social cues and behaviors. Breazeal's approach is directly inspired by the dynamics of the parent-child caregiving relationship, aiming to replicate this bond between humans and robots (Breazeal & Velásquez, 1998). Jibo was, in fact, marketed with the tag line that it was "the world's first family robot" (Bielski, 2014).

Considering that caregiving offers humans their initial exposure to a feeling of attachment (Bowlby, 1983) and security (Sroufe et al., 1993) via parent-infant interactions, it is plausible that the caregiving role of robots will also play a substantial role in fostering psychologically meaningful interactions between humans and robots. In the context of elderly care, social robots can fulfill roles that resonate with these attachment needs, particularly when human interaction is limited.

Social robots designed with the capability—or believable impression of the capability—to respond to human emotions and engage in social behaviors can foster a sense of connection and reciprocity. The consistency of care and interaction provided by a robot, unlike human caregivers who may change or be unavailable at times, may also enhance feelings of trust, attachment, and reliance. Elderly users might then come to rely on the robot for social interaction, assistance with daily tasks, or reminders for medication, further strengthening an emotional bond (Kidd & Breazeal, 2008). However, with every emotional bond comes the potential of emotional vulnerability such as risk of loss, such as if the robot was to be broken or replaced with a noticeably different model.

Yet while older individuals play a role as a significant focus of the advancement of social robotics, they are often overlooked during the design phase, with their specific concerns left unaddressed. Instead, the design process tends to focus on broader shifts in the rapidly increasing aging populations and related healthcare demands, attributing general needs to older individuals without acknowledging their unique perspectives. This approach can rely on stereotypes, portraying older people as isolated, fragile, and reliant on robotic aid. Research cannot rely on expectations of the experiences of aging via empathy and prior existing research, which is still limited. Building hypotheses about aging populations and social robots on empathy alone are thought exercises that are interesting to the researchers but not necessarily salient in everyday life.

There is a critical need to reassess and redefine perceptions of older individuals to accurately represent their diversity and needs. Further *in situ* research into older adults as users of social robotics is essential to better cater to their real specific needs.

FAMILY MAINTENANCE WITH A ROBOT

Pivotal to the central moral quandaries of *Robot & Frank* are the differing opinions of Frank's adult children about having a live-in robot assistant for their father. Hunter, Frank's son, initiates the idea of using the robot to care for Frank because he is concerned about his father's well-being and sees the robot as a practical solution that can help ease the burden of caregiving. Hunter believes that the robot will provide consistent assistance and surveillance and will ultimately maintain Frank's autonomy as long as possible and thus improve Frank's quality of life. However, Madison, Frank's daughter, is opposed to the idea of the robot and feels uncomfortable about a machine caring for her father, considering it to be a cold and impersonal approach. Madison worries that the robot will replace the human touch in her father's care and expresses a desire to handle the situation of her aging father by herself.

The term *sandwich generation* was first coined by social worker Dorothy A. Miller in 1981 to describe middle-aged adults who are simultaneously responsible for caring for their own children and their aging parents. This phenomenon has gained increased attention due to demographic and economic shifts, leading to significant implications for those involved. According to the Pew Research Center (Parker & Patten, 2013), nearly half of Americans in their 1940s and 1950s find themselves in this situation. The prevalence of the *sandwich generation* is partly due to increased life expectancy and delayed childbearing, causing many adults to find themselves in caregiving roles at both ends of the age spectrum.

The demands placed on the sandwich generation can lead to considerable stress and burnout. Frack and Chapman (2021) highlight that these family members often experience time constraints, emotional exhaustion, and financial strain due to the dual caregiving responsibilities. Balancing work commitments and personal life adds to these challenges, making the situation particularly burdensome. Research by Parker and Patton (2013) indicates that caregivers in the *sandwich generation* are at a higher risk of mental health issues such as depression and anxiety due to the stressors of caregiving. The dual role often requires sacrifices in personal leisure time, resulting in less self-care and higher susceptibility to health problems.

Anticipating a need for in-home robots—especially for eldercare—Honda developed Advanced Step in Innovative Mobility (ASIMO) (Shigemi, 2018). ASIMO was first introduced in 2000 and designed to assist people in everyday environments, its lightweight and compact design making it potentially suitable for home use (Eaton, 2015). The robot could recognize human voices and gestures, navigate obstacles, and perform tasks such as carrying trays and pushing carts. At the time, ASIMO represented a significant achievement in robotics. Its history traced back to the 1980s, when Honda started its journey toward creating a robot that could operate safely in human environments, establishing a humanoid robotics program and producing several prototypes leading up to the final design of ASIMO. The first experimental model, E0, was developed in 1986 and was followed by a series of improved prototypes known as E-series and P-series over the next decade (Eaton, 2015; Shigami, 2018).

In its origin story, as told by Honda, ASIMO was designed with careful considerations for integrating into a domestic, everyday setting (Honda Robotics, *"What We Learned from ASIMO,"* 2024):

> In order to create robots which can operate in a living environment with people, Honda pursued various initiatives to ensure the safety of nearby people, including the development of technology which will absorb the impact of a collision between the robot and a person.

Still, we were thinking, "What if there were children at the spot where our robot fell over?" What we realized through our ASIMO research was that in order to let robots co-exist with people in a real-world situation, we must assume many more scenarios and be even more conscious about the safety of people.

Regarding the robot's overall aesthetic, ASIMO North American project lead Jeffrey Smith said, "ASIMO's good looks are deliberate," and clarified that a humanlike appearance is "key to ASIMO's acceptance in society" (Ulanoff, 2003). At the time, Smith explained that he envisioned ASIMO as a future assistant to the elderly and disabled people to improve their life experiences.

Functionally, among other things, the semiautonomous robot could take instructions from users and provide them with services including home surveillance, communication assistance, and light household chores, and was adept at recognizing multiple moving objects and faces. The 2011 ASIMO model was 130 cm tall, weighed 48 kg, and designed at this height specifically to aid people in wheelchairs (Roy, 2021). Additionally, Honda's research determined that the optimal height for a mobility assistant robot ranges from 120 cm to the average adult height, making it better suited for tasks like operating doorknobs and light switches.

In 2004, *The Guardian* reported Honda's hopes for ASIMO's technology would continue to advance with the goal of integrating into domestic life. "As the next five years go by and we refine it, then it's going to be able to do voice-activated commands. Then you could use it for old people that were infirm, perhaps" (Jha, 2004).

However, technological limitations and high development costs meant that ASIMO remained primarily a research and demonstration tool rather than becoming deployed as a widespread household assistant ("Honda walking assist device," 2019; Sakagami et al., 2002). Yet, over the years, Honda continued to refine ASIMO, enhancing its ability to walk, climb stairs, and understand human interactions. While not widely deployed in homes, ASIMO played a critical role in advancing robotics research and demonstrated the potential for humanoid robots to interact safely and effectively in human environments (Hirai et al., 1998).

In 2002, ASIMO began its role at Tokyo's Miraikan Museum as an interpreter and science communicator, and exhibition commentator. During its tenure at the Miraiken, the robot greeted UK's Prince Charles during a Royal visit and performed a seven-minute dance routine in his honor

© 2002 American Honda Motor Co., Inc. honda.com

We're building a dream, one robot at a time.

The dream was simple. Design a robot that, one day, could duplicate the complexities of human motion and actually help people. An easy task? Hardly. But after more than 15 years of research and development, the result is ASIMO, an advanced robot with unprecedented human-like abilities. ASIMO walks forward and backward, turns corners, and goes up and down stairs with ease. All with a remarkable sense of strength and balance.

The future of this exciting technology is even more promising. ASIMO has the potential to respond to simple voice commands, recognize faces, carry loads and even push wheeled objects. This means that, one day, ASIMO could be quite useful in some very important tasks. Like assisting the elderly, and even helping with household chores. In essence, ASIMO might serve as another set of eyes, ears and legs for all kinds of people in need.

All of this represents the steps we're taking to develop products that make our world a better place. And in ASIMO's case, it's a giant step in the right direction.

HONDA
The Power of Dreams

FIGURE 2.1 (Honda, 2002). "Honda, the power of dreams" print ad campaign. PRIVATE COLLECTION: Julie Carpenter.

(Alderson, 2008). By the end of 2021, ASIMO had conducted 15,466 public demonstrations in the museum. To commemorate the special event, the museum threw an event to celebrate its accomplishments called, "Thank you, ASIMO! Congratulations on graduating from Miraikan" (Miraikan Museum, 2022).

Throughout the 2000s, ASIMO would continue to make appearances globally to demonstrate its technological capabilities, garner interest in science and technology, and generally entertain audiences. Honda announced that in 2003, ASIMO would go on a 15-month nationwide educational tour, entirely funded by Honda, targeting students from grades five through high school, as an initiative aiming to inspire children about technology (Ulanoff).

ASIMO was also significantly featured in a 2005 show titled *Say "Hello" to Honda's ASIMO* at Disneyland's Innoventions attraction, located in the Tomorrowland area of the park (Honda News, 2005). This installation marked the only permanent display of ASIMO in North America until the closure of Innoventions in April 2015 (White, 2007).

In January 2010, Honda premiered the *Living With Robots* documentary at the Sundance Film Festival, a film highlighting everyday human interactions with robots (Honda News). The following year, ASIMO was showcased at the FIRST Championship in St. Louis, Missouri, to inspire students to study math, science, and engineering (Volkmann, 2011).

In 2018, Honda announced it would stop the development of ASIMO models in favor of focusing on other humanoid robots for eldercare and disaster relief (Ackerman, 2018). They had already begun shifting away from ASIMO's development since July 2013, when Honda instead focused on a disaster response robot, in part built using technology gained from ASIMO. Even though ASIMO has been retired, the robot foreshadows what the future likely holds, and that is versions of robot companions and assistants in homes.

The concept of a robot in the home as a reality is now in the global consciousness, largely due in part to via Honda's decades-long ASIMO public relations campaign. Furthermore, ASIMO provided a social robot mental model for people's imaginations and expectations in terms of its capabilities and limitations.

PERCEPTION OF ANIMATION AND INTENT AS INTELLIGENCE

Robot movement implies intent, an internal state of intelligence and level of autonomy to its observers. Darling (2016) has highlighted the social implications of humans' perceptions of autonomous motion by robots, particularly when these machines are meant to mimic life through movement, and Breazeal (2003) posits that social robots that can express and perceive emotions through movement can engage in more meaningful interactions with humans. The challenges lie in designing social robots that can use movement to carry out their tasks and facilitate positive interactions with people. However, how to do those things without misleading users about the robot's capabilities or fostering inappropriate dependencies is problematic by the nature of the robot's role as caregiver in a situation like in-home eldercare which requires—at a minimum—gaining user trust.

The way humans perceive movement in social robots significantly influences their attributions of intelligence to it. *Animation* refers to the dynamic behavior of robots, including facial expressions, gestures, and vocalizations. Animated robots can convey emotions, intentions, and social cues, enriching the interaction experience. For example, robots equipped with expressive faces can display a range of emotions, allowing for more nuanced communication with humans. Animation also enables robots to adapt to different situations and convey empathy, fostering deeper connections with users, thus, movement, motion, and human-robot space are critical aspects of human-robot interaction. Furthermore, smooth, and fluid movements from a robot may be associated with a sense of gracefulness and competent engineering.[1] In contrast, a robot that has jerky or erratic motions may be perceived as uncoordinated, malfunctioning, or threatening. Studies have shown that human-like gestures and movements can enhance social acceptance and engagement with robots, particularly in collaborative tasks (MacDorman & Ishiguro, 2006). Additionally, synchronized movements and shared attention between robots and humans can facilitate communication and cooperation (Policastro et al., 2009).

Hoffman and Ju (2014) explored how dynamic and responsive movements in robots could lead to higher attributions of intelligence by human observers. They argued that robots with fluid and purposeful movements are often perceived as being more intelligent than those with mechanical or predictable movements. This perception stems from humans' natural tendency to associate movement patterns with the presence of life and cognitive processing.

Movement affects perceptions not only of intelligence but also of intent. The ability of a robot to move in a way that appears intentional, such as turning toward a sound or reaching out to an object, can give the impression that the robot has desires or goals, thereby enhancing its social presence. Saerbeck and Bartneck (2010) demonstrated that robots designed with non-verbal cues, such as expressive movements, could elicit specific emotional responses from humans. These responses were significantly influenced by the perceived intentionality and appropriateness of the robot's movements, affecting trust and likability.

Daniel Dennett's (1989) discussions on robots, embodiment, movement, and intentionality revolve around the idea that intentionality—the capacity of the mind to be directed toward an object or to have thoughts about something—is not exclusive to biological entities but can also be attributed to robots under certain conditions. Dennett argues that for robots to exhibit intentionality, they must have a form of embodiment that allows them to interact with and move within their environment. This interaction enables them to gather and respond to sensory information, thereby demonstrating a rudimentary form of intentionality. Dennett emphasizes that intentionality arises from the functional organization of a system, rather than its material composition.

EMBODIMENT, APPEARANCE, MORPHOLOGY, AND TOUCH

In the movie *Robot & Frank*, the titular character of Robot, embodied in costume by actor Rachael Ma, is diminutive in height compared to Frank, a shiny white biped robot that moves with distinct mechanicalness. As a character, Robot's physicality was inspired by the design of Honda's very real ASIMO robot.

FIGURE 2.2   (2012). Actors Frank Langella (as Frank) and Rachael Ma (as Robot).

Source: Samuel Goldwyn Films

Robot embodiment, the physical form and appearance of robots, plays a pivotal role in human-robot interaction and social robotics. *Embodiment* in robots refers to their physical manifestation, encompassing factors such as appearance, movement, and interaction capabilities. The embodied character of the robot's body includes its physical characteristics (e.g., head, face, torso, or legs) and its capability at learning by gaining information through sensory input received via various ways, such as cameras and microphones. As a whole, a robot's embodiment significantly influences human perceptions and behaviors, shaping the quality and nature of the human-robot interaction (Goetz et al., 2003; Kiesler et al., 2008).

The significance of embodiment in social robots unveils a profound ontological inquiry into human-robot interactions. Embodiment delineates not just the physical presence of the robot but the existential implications it engenders within the intersubjective realm. Through this lens, the body of a social robot becomes a locus of meaning and interaction, one that mediates an existential encounter between human consciousness and technological otherness. Marshall McLuhan, media theorist, viewed technology as extensions of human faculties that shape and influence the way people perceive and interact with the world (1964). One of McLuhan's central concepts is the idea that "the medium is the message," suggesting that the form and characteristics of a medium have a profound impact on society, independent of its content (McLuhan, 1964). In the case of social robots, their physical form, interactive capabilities, and communicative functions serve as mediums that shape human experiences and relationships.

The embodiment of social robots introduces a dynamic interplay between human agency and the artificial agency of the robot. This relationship elucidates questions of authenticity and autonomy, as humans navigate their encounters with embodied artificial agents like robots and negotiate the boundaries of the self and other in the process. Moreover, the gaze of the robot, whether programmed or perceived, invokes a dialectic of recognition and objectification, mirroring Sartre's concept of the *le regard* as a tool for subject-object relations. Thus, the embodiment of social robots imbues them with a social presence that elicits existential responses from humans, shaping the dynamics of intersubjective encounters.

Using factors like robot morphology and movement combined, people identify robots as potential social others and can perceive their movement as indicators of lifelike and intelligent categorizations, even when they understand it is a nonliving entity or a robot (Bernotat et al., 2021; Liberman-Pincu et al., 2023; Carpenter, 2016). Critics to the purposeful anthropormorphic design of robots may be fighting against a human instinct to interpret the combined factors of a robot's role, movement, embodiment, communication, and cultural priming with stories that model expectations of human-robot interactions. Combined, these are all powerful social indicators of something that is life*like* and, therefore, to be understood and regarded in a way that needs an appropriate model of interaction. Interaction with robots is still a new phenomenon for many people, and people are individually and collectively negotiating how to interact with robots in a situation that indicates a social paradigm.

Thus, the embodiment of social robots raises ethical considerations reminiscent of existential ethics. The design and deployment of these robots implicate questions of authenticity, freedom, and responsibility, as humans grapple with their roles as creators and interactors within technologically mediated social landscapes. Embodied robots with lifelike features and behaviors evoke emotional responses and social cues akin to human-human interaction, theoretically facilitating more natural and intuitive communication (Breazeal, 2002). Embodiment also effects *acceptance* or the user's perception that the robot *should be* integrated into the environment, which in the case of eldercare could be a private room, home, hospice, or healthcare facility. Acceptance also includes a belief that the robot is useful, their interactions are appropriate, helpful, and adaptable to all of the people around them. Therefore, the significance of robot embodiment lies in its ability to bridge the gap between humans and machines, enabling robots to function effectively as social agents in various domains, including eldercare.

Equally important, cultural factors also influence how the shape and size of robots are perceived and accepted. For robots designed for socialness, the way humans perceive their shape and size is crucial for several reasons, particularly in how these perceptions influence interaction, acceptance, and the effectiveness of robots in social roles. The shape and size of a robot can influence how lifelike, relatable, or *safe* it appears, impacting the level of empathy and engagement humans feel toward it. A robot's size, for instance, must be appropriate for its operational environment to minimize intimidation or physical risk to humans; the shape and size of a robot affect its functionality and the efficiency with which it performs tasks, which in turn affects human perceptions of the robot's usefulness (Chatzoglou et al., 2023). Robots with the relative size and shape of a humanlike body in a robot can also, theoretically, move efficiently in spaces made for people, such as private homes, retail stores, and other everyday spaces. Consequently, the physical design of a social robot, including its size, should match its functionality—including socialness—to ensure effective interaction with humans.

Additionally, interacting with robots can shed light on how humans perceive their bodies and navigate a sense of self (Ehrsson, 2007), further implying there can be potential for robots in both physical and cognitive therapies. Jacques Lacan emphasizes the significant moment when a young child first encounters their mirror image, describing it as a fundamental precursor to identity formation. In Lacanian psychoanalytic theory, this pre-conscious recognition of oneself in the mirror represents an initial stage in the process of identification with an other, ultimately leading to the development of the subject (2004). This early self-image is intricately linked to the formation of the subject, existing in a complex relationship characterized by asymptotic tendencies. Within this framework, the distorted or even exaggerated mirror-like reflection of social robots serves as a fictional representation of something living, symbolizing not only the enduring presence of the self but also its representation as a distortion, as portrayed through the robot's asymmetrical representation of otherness.

The significance of tactileness and touch for humans and social robots also unravels profound existential inquiries into the nature of being and relationality. As humans, the lived experience of the body is a site of consciousness and engagement with the world (Ehrsson et al., 2007). Touch, as a sensory modality, embodies the existential encounter between the self and other, illuminating the complexities of human existence within relational contexts. Similarly, for social robots, tactileness holds significance as it mediates their interactions with humans, shaping the dynamics of intersubjective engagement.

The tactile exchange between humans and social robots parallels a sort of human-human mutual recognition, evoking questions of authenticity and relationality within technologically mediated encounters. The touch of the robot becomes a conduit for reflection, inviting humans to confront their own subjectivity and the artificiality of the other. The tactile experience of robot embodiment serves as a poignant reminder of the fragile nature of human existence and the complexities of relationality in an increasingly technologically mediated world.

UNCANNY VALLEY AND CHANGING EXPECTATIONS OF OTHERNESS

The *Uncanny Valley* theory was introduced by Japanese engineer Masahiro Mori in 1970. This theory posits that as robots (or any entity) become more humanlike in their appearance and movement, the emotional response from human observers moves from empathy to revulsion, before returning to a more positive response as the *affinity* (or resemblance) to humans becomes virtually indistinguishable. Mori described this dip in emotional response as the Uncanny Valley. The concept has significantly influenced both theoretical and practical aspects of robotics, human-computer interaction, and even entertainment industries. However, it has also some faced criticism regarding its empirical basis, universality, and applicability across different cultures.

In 2012, Mori ruminated on his inspiration for his original concept (Kageki),

> Since I was a child, I have never liked looking at wax figures. They looked somewhat creepy to me. At that time, electronic prosthetic hands were being developed, and they triggered in me the same kind of sensation. These experiences had made me start thinking about robots in general, which led me to write that essay. The uncanny valley was my intuition.

In the same article (ibid.), Mori went on,

> You can't define a robot. It's the same as trying to define Mt. Fuji. If a steep hill suddenly protrudes from the flatland, you can draw a line to show where the mountain starts, but Mt. Fuji becomes higher so gradually that you can't draw a line. Robots are like Mt. Fuji. It's hard to separate what is a robot from what is not. ASIMO is so near the peak, anyone can easily call it a robot. But what about a dishwasher? It can automatically wash dishes, so you might call it a robot. The line is blurry.

Still, designers and engineers have used the Uncanny Valley as a cautionary principle to avoid creating robots that fall into the Uncanny Valley, aiming instead for designs that either deliberately avoid humanlikeness or attempt to achieve a highly convincing humanlike appearance and behavior. The entertainment industry, particularly in film, animation, and video game design, has been significantly influenced by Mori's concept. Filmmakers and game designers have strived to create characters that are either stylized to avoid the uncanny valley or so realistically human-like that they do not evoke a sense of unease in viewers and players

Now, as robotics has evolved beyond the realm of theoretical, Mori's theory needs to be situated culturally to account for responses to humanlike robots which can vary significantly. The Uncanny Valley places primacy on a person's perception based on a robot's appearance and movement regardless of the use case situation, the robot's other abilities (e.g., voice), cultural context, whether other people are present, or accounting for the user's familiarity with the robot *over time*. In the original Uncanny Valley theoretical framework, examples of movement and embodiment of entities are given as suggested data points on the graph (e.g., a still stuffed animal; a moving industrial robot). This premise does not address that affinity with any object that is initially viewed as *uncanny* may increase as someone is exposed to it over time, and thus, that entity becomes ordinary or normalized. Thus, the Uncanny Valley also does not consider cultural priming, or previous encounters with robots in real life, and news or other factual descriptions, artistic representations—including storytelling—that may have set peoples' expectations about robots.

Especially in this stage of general robot discovery as social actors for most people, *expectations* of robot interactions are what informs initial behaviors with social robots (Carpenter, 2016; Heerink et al., 2010; Kwon et al., 2016; Saari et al., 2022). These expectations influence and model how people approach, communicate with, and respond to robots, impacting the overall effectiveness and nature of the interaction. Expectations about HRI are shaped by a variety of factors, including prior experiences with technology, cultural background, media portrayals of robots, and individual predispositions toward technology (Carpenter, 2016; Horstmann & Krämer, 2020). The position of the human gaze highlights this role of preconceived notions and media representations in shaping both design *and* use expectations, which can lead to either overly optimistic or unduly fearful anticipations of robot capabilities. Additionally, *expectancy theory* suggests that if people expect robots to be intelligent and capable of social interaction, they are more likely to engage in complex communications and attribute social qualities to the robots (Heerink et al., 2010). Conversely, if expectations are low, interactions may be more cautious or limited, potentially inhibiting the development of meaningful connections with the robot. To ethically manage these expectations for peoples' adaptation to interacting with real robots, designers and researchers must manage users' expectations by making it normal to provide realistic information about a robot's capabilities, and to design robots that can meet these expectations so as not to disappoint (Eyssel et al., 2012) users.

Research suggests heavily that humans attribute humanlike qualities to robots with anthropomorphic features, such as humanoid robots or those with expressive faces (Breazeal et al., 2016). For example, these robots may be perceived as approachable and trustworthy, leading to more positive social interactions (MacDorman & Ishiguro, 2006; Song et al., 2021). Yet, robots with nonhumanoid shapes may also evoke feelings of social otherness responses depending on the situated factors of a robot's use (e.g., to serve), context (e.g., the home), and its overall shape (e.g., petlike morphology), and or movement. For example, the home vacuum nonhumanoid robot Roombas have been widely documented as being zoomorphized or anthropomorphized by their owners (iRobot Blog, "*The Top 50 Most Popular Roomba Names,*" 2012; Sung et al., 2007; Fink et al., 2012), while a nonhumanoid robot in a factory setting may be regarded more as a functional tool (Evjemo et al., 2020).

Cynthia Breazeal (2003) discusses the design of robots capable of expressing and recognizing emotions, suggesting that such capabilities can facilitate empathetic connections *between* humans and robots. The proposition then is that social robots can evoke empathy from humans, positioning them as others, which are then capable of entering into a social relationship with the user. However, this also raises ethical questions about the nature of the relationship and the obligations humans might feel toward robotic others.

By acknowledging and engaging with the subjective experiences of others, individuals cultivate *empathy* as a means of fostering mutual understanding, compassion, and relationality. Phenomenology conceptualizes *empathy* as a fundamental aspect of human experience, emphasizing the interconnectedness of subjective consciousness and the intersubjective world. The experience of one's own body as one's own subjectivity is applied to the experience of another's body, which, through apperception, is constituted as another subjectivity. One can thus recognize the other's intentions, emotions, and behaviors. Empirical research demonstrating that robots with a human bodily form evoke social-emotional reactions from humans is exemplified in the study conducted by Riek and her team

(2009), whose findings claim that "the greater the resemblance of a robot to a human, the stronger the empathetic response it elicits."

In the context of a human-social robot interaction, this encounter raises questions not just about demarcation and physical boundaries but about empathy, moral consideration, and the capacity to recognize robots as a quasi-other. Turkle (2011) explores how individuals perceive robots not merely as tools but as entities that occupy a unique space between the object and subject. This perception is influenced by the robot's humanlike or lifelike qualities, which can evoke feelings of social presence and elicit social responses from humans. The anthropomorphic and zoomorphic designs of robots further complicate the self-other distinction, as they encourage users to attribute human or animal-like qualities to machines.

However, this phenomenological perspective necessitates a focus on both the *subjectivity* of the other person and the *intersubjective* engagement between individuals. According to Husserl (Carman, 1999), this process involves a form of apperception that is built upon one's own lived experiences, particularly those related to one's own body. The concept of one's *lived body* primarily manifests itself through the possibilities it offers for action in the world. For instance, it enables individuals to reach out and grasp objects, while also allowing for the potential to change one's perspective. This ability is crucial for distinguishing between different entities, as individuals can move around an object, perceive new aspects of it, and yet maintain the recognition that it remains the same object observed moments earlier (this concept is often described as making the "absent present and the present absent," ensuring its identity). Furthermore, individuals experience their body dually: as an object (e.g., one's ability to touch one's own hand) and as a manifestation of their own subjectivity (e.g., one's experience of being touched).

Moreover, the lived body is central to being-in-the-world, as it is through bodies that living organisms—and robots—perceive, interact with, and understand their environment. This view emphasizes that the body is not just a biological organism but a medium of perception and intentionality (Husserl, 1962; Gallagher, 2006; Merleau-Ponty, 1962). Thus, bodily movements and sensations are fundamental to how people experience and make sense of the world. The lived—or in the case of robots, inhabited—body is thus a bridge between the self and the world, shaping and being shaped by interactions with it.

TAKESHI KOSHIISHI

Honda Japan, Designer of ASIMO, Chief Designer, Innovation Studio, Creative Solution Center Honda Motor Co., Ltd.

Takeshi Koshiishi joined Honda in 1993. In addition to robot design, such as his masterpiece ASIMO, Koshiishi has been passionately involved in pioneering designs such as the Honda Walking Assist Devices, a series of robotic exoskeletons. Research on the Walking Assist Devices began in 1999, utilizing the advancements from Honda's extensive study of human walking, which also contributed to the development of the eventual design of the nimble and bipedal humanoid robot ASIMO (Honda in America, 2019). He also has been involved in the design of Predicting Action of the Human (P.A.T.H.) Bot, Mobile Power Pack Charge & Supply Concept model, Honda's hot-air balloons, and racing wheelchairs. During our email correspondence for this book, Koshiishi conveyed a sincere passion for designing assistive technologies that are not only efficient but pleasing for people to interact with.

Throughout the 2000s, ASIMO was clearly influential as a model of social *robotness*. Because of this, I wanted to know more about the creative influences and inspirations behind its successful design that produced such an innovative robot that charmed audiences globally. Koshiishi and I communicated via email directly in English, but this interview was conducted more formally and translated by Honda staff. His sincere love of science fiction and storytelling shine through in the medium of robot design and his influence on ASIMO, which, in turn, inspired screenwriter Christopher Ford and countless others to tell new stories.

I would like to know if Koshiishi-san has stories he thinks of as influential or significant to him and if so, what they are and how he feels they influence or inspire his design work or broader philosophies. His response does not necessarily have to be ASIMO- or Honda-specific.

The people who have influenced my work, and who I respect are Soichiro Honda, Osamu Tezuka, and Walt Disney.

I loved riding my bike. I could move farther, faster than walking. Naturally, I became interested in motorcycles, starting with 50cc and gradually moving up to larger displacements.

The exhilaration of riding up hills that would be hard for a bicycle. The wind, the acceleration, the sound of the engine fascinated me, taking me to unknown places faster. It was not just a vehicle, it was like having wings.

After riding a wing-logoed, Honda motorcycle, I learned about Soichiro Honda.

He was a brilliant engineer, designer, and artist as well as a business owner. He valued customers and dreams the most.

He passed away shortly before I joined Honda, and I was sorry that I could not meet him. I wonder what he would have said if he had seen ASIMO?

When I was a child, I wanted to be a doctor or a cartoonist. In the end, I didn't become either, but I believe that my current design work has a little of both in it.

Osamu Tezuka was a doctor and a cartoonist, and his works cover a wide range of subjects, including robots, mythology, medicine, and religion, all of which are so wonderful that he is known and respected in Japan as the God of Manga.

Astro Boy is a wonderful series, and has influenced many people. Although I am not of the Astro Boy generation, I bought and studied all the manga volumes when I joined the Honda Robot Project. The essence of Astro Boy is not his achievements with his seven powers, but the serious depiction of how people and robots, and robots and robots, relate to each other.

My favorite of Tezuka's work is the Phoenix series, and the Resurrection volume is a masterpiece that examines people and robots even more deeply after Astro Boy. The story of a robot called Robita, with the heart of a human, was very thought-provoking.

I think it is unfortunate that robots, made in the image of humans, inherits human contradictions and even human destiny. Only human beings should have that burden.

People sometimes say that the Japanese have a high affinity for robots, largely due to the influence of Astro Boy. Seeing the smiles on people's faces around the world as they watch an ASIMO demonstration, I believe this is a universal phenomenon, not limited to the Japanese.

I learned a lot about Walt Disney from reading his biography and through watching documentaries. "Disney Animation The Illusion of Life," a book written by a Disney animator, was helpful in my work. I visited Walt Disney Studios in Burbank when I was on a business trip to set up an ASIMO mock-up at Disneyland Autopia in Anaheim. This was the very place where Walt Disney himself had created many of his works, and I felt as if his spirit resided there.

His words, "Disneyland will never be completed. As long as there is imagination left in this world, it will continue to grow," resonated with me.

When we were developing the first ASIMO, I presented it directly to then-president Yoshino, and received his approval for the design. He smiled while listening to my presentation, in which a cheeky and young I said, "The word 'robot' comes from RUR (Rossum's Universal Robot), but the earliest record of robotic beings is 2,800 years ago, and it is the dream of mankind."

I later learned Yoshino also studied robots deeply, including its philosophy. Thinking back on it now, I am embarrassed. I was like Son Goku, who makes himself look great with the palm of Buddha. (I thought I knew it all, that I was the expert, but all along he knew better.)

I will never forget his words, "Listen, the energy of dreams is inexhaustible and unlimited. You just have to think that way."

When Koshiishi-san was a child, did he have favorite stories that emotionally resonated with him, and what were they? I am interested if/how he connects those stories or ideas to his creative or professional path today.

When I was a child, I often stayed at home because of my illness, reading fantasy science picture books. Among them, "What If the World" left a strong impression on me. I was thrilled reading the text and seeing the illustrations on themes such as "What if the oceans ran out of water," "What if dolphins could talk to humans," and "What if there was a medicine that cured death?"

Even now, I still sometimes ponder "What if . . . ?" This time is precious to me. The harder I think, the closer I can get to closer to the ideal.

I was a huge fan of Star Wars since I was a kid. I went with my family to see "Star Wars: The Empire Strikes Back" (my favorite in the series) at a theater. It was very crowded and we had to stand, but I didn't get tired at all. I was shocked when Luke's hand was cut off by Vader's lightsaber after a fierce battle between them. I felt very gloomy, thinking that children should not be allowed to see this. But later in the movie, a medical robot gave Luke a new, robotic hand. I was relieved because it all seemed real, even though it was all made up. The evolution of technology is a wonderful thing. Not only can we restore body functions, but we can also augment them further. Darth Vader is a cyborg.

I think it was around this time that I became interested in technology that could improve people's abilities and products that could contribute to medical care.

I didn't think about these things all the time, but obviously the fragments of my memories from that time are helping me in my work today.

In "The Force Awakens," reclusive Luke's robotic hand is uncared for, fleshless.

The robotic hand is also used symbolically in "The Terminator." In "2001: A Space Odyssey," the caveman's bone, used as a weapon, it triumphantly thrown into the air, morphing into a spaceship. I understood from watching these movies, is that a key to human evolution, rivalling bipedalism, is the dexterity of the hand, which allows the thumb to be used opposite the four fingers.

Androids, humanoids, replicants, artificial bodies, the names may vary, but beings that reflect people have been depicted repeatedly in stories since ancient times, so much so that they can be said to be a dream of humankind.

I think the fact that many of these stories are about enemies of humankind may be due to our instincts as living creatures. It is unthinkable that we will be attacked and preyed upon by a ferocious beast in our daily lives. We have reached the top of the food chain, and our lives are no longer in danger. Still, if there are two circles on a wall, we see them as eyes or faces, and we are subconsciously on guard.

On the other hand, friendly robots are also depicted. Just like a mirror, everything depends on how we are.

As Professor Mori's theory of the Uncanny Valley suggests, there is indeed a deep valley. As a designer, I understood this intuitively. Airplanes do not flap their wings like birds. In the same way, we need to consider how closely we should reproduce the human form according to purpose. The essence is not the form, but whether others can feel the [humanoid's] heart and approve of it.

I personally believe "humanoid" has an element of discrimination hidden away. The term is generally used to describe a bipedal form with a head, arms, and legs, but there are many people in the world who wouldn't fit this description.

Also, the word "robot" originates from servitude, not a good word, but it is so widespread now, there is no word to replace it. I hope we can find a new word.

What are more recent stories that emotionally resonate with Koshiishi-san as an adult that influence his creative, professional, or personal path? Again, "stories" do not have to be about technology (as in the example I gave from Dr. Mori-san). Other examples can include bedtime stories, family stories, mentor anecdotes, movies, books, manga, comic books, etc.

In art school, I was exposed to user interface design, a field in which my former teacher was a leading expert. I was given the opportunity to exhibit my work in Berlin three years after the fall of the Wall, and I also visited Bauhaus, which was very inspiring.

I was no outstanding student, but I felt a particular sympathy for the relationship between people and artifacts and the idea of a time axis. The concepts of movement, sound, and time were areas of design that could not be expressed on paper, and a PC was essential.

At the time, use of PCs in design was not as common. I learned how to use 2D and video editing software on the Mac, and 3D software Alias on an SGI workstation.

In robot design, using 3D software is indispensable to create beautiful shapes while conforming to complex functional movements, and my experience as a student has been very useful in my current design work.

The Mac, created by Steve Jobs, is a creative tool unmatched by any PC before it, and above all, it is simple and beautiful. His words, "Technology dies out after a short period of time, but the stories it creates are passed on for decades, even centuries," stuck with me. People were moved to tears when the satellite Hayabusa finally returned from its difficult mission on the asteroid Itokawa, and burned up entering the atmosphere, leaving its capsule behind.

There is a Japanese custom called *Hari Kuyo*, in which needles are placed in soft tofu and sutras are read to express gratitude for broken needles that are no longer useful. I believe that it is most important for people to be thankful for things and the stories they tell.

I have watched many science fiction movies, not to mention *Star Wars*. I was deeply moved by the ending of *Blade Runner*, depicting Syd Mead's worldview, in which a replicant felt the importance of life and decided not to kill the protagonist. (I wonder what would happen if I were to take the Voight-Kampff test?)

The robot in *Interstellar* has a square design that seems to be inspired by the Monolith, and the way it walks is innovative. It also has a sense of humor. *Westworld* is a cruel story, but it was very well made. I like Western movies, so I was drawn into the story.

I recently watched *The Peripheral*. Its depiction of movement in time axis in addition to location is interesting, and I am also interested in parallel worlds. The way a Peripheral, which has the same appearance as a human, switches between autonomous AI mode and remote mode with the operator's consciousness is brilliant. Other robots also appear, some faceless and others with faces displayed on a mesh screen. I look forward to Season 2.

I watch a lot of movies, and believe that ASIMO's influence on movies is not insignificant.

In the past, robots looked shiny and mechanical, but since the appearance of ASIMO, white robots seem to have been depicted more often.

(Originally, RUR's original robots were not mechanical, but artificial humanoid robots made by bio-synthesis.) ASIMO walks more smoothly than C3PO, which has a human inside and is acting as a robot, and this transcends the world of imagination to reality. Also, R2D2 and C3PO are a great duo, but I sense George Lucas' intention in that R2D2, which does not take human form, is smarter than C3PO.

Robot & Frank was not shown in many theaters, but I went to see it right away at a small movie theater. It was a very hot day in the middle of summer. I enjoyed the story of old Frank, who remembers his younger self through his wrongdoings with the robot, and how the robot becomes a helper of his heart.

The design and movement of robots based on ASIMO show the hard work that goes into overcoming difficulties of having an actor inside. The contrast of the helper robot with C3PO, where a person is acting like a robot, is interesting.

In the area where I live, Noh theater performed by fire is held one night each summer in a park, and I used to go to see it every year before the Corona disaster. The performance begins in the evening, and as the crows cawed and it gradually grows darker, I am fascinated by the ethereal and otherworldly atmosphere of the Noh play performed on a stage lit by firewood. The eyes and mouth of the Noh mask is immutable, but the angles and movement gives the mask a sense of facial expression.

ASIMO's face is covered with a smoke shield to hide the facial details. This is intended to allow the onlooker to imagine ASIMO's expression. In fact, in early development, we released a version of ASIMO with a clear face shield, its facial features clearly visible. Although it was popular in its own right, it did not gain a lot of support. We believe this was because it inhibited people's imagination. The murals of Mary and the saints in the Rosary Chapel in France, which Henri Matisse was involved in, do not depict faces. Matisse says this is because anyone can paint them in their mind.

During design development of the first ASIMO, a "design clinic" was held in which participants were shown mock-ups of two different designs and asked for their impressions.

We discussed the possibility of replacing the head of each design. Being young and inconsiderate, I lifted the robot's head and removed it while everyone was watching. The audience gasped.

A mock-up is nothing more than a lifeless object, but the onlooker does not see it that way. I realized that whether there is a soul or not, is decided by the person who sees and feels it. Human sensitivity is the most important thing.

Later, when the completed ASIMO walked in front of people on stage, it clearly looked like a it had a soul.

A boy with progeria syndrome in India liked ASIMO so much, he wanted to meet ASIMO.

He had three dreams: To meet ASIMO, to go to Disneyland in California, and the third, he would tell us when he grew up. In 2016, the local Honda office invited him and his family to the Delhi Motor Show, where he met ASIMO and was thrilled. I thought it was a wonderful story.

In 2017, when Disneyland Autopia was completed with ASIMO installed, I suggested that we invite the Indian boy. He had already passed away. I hoped ASIMO had given his life even a little bit of joy.

Through ASIMO, I learned awesome and wonderful human beings are, and I worked on ASIMO selflessly.

I have heard that some young people have been inspired to become engineers thanks to ASIMO.

I believe that ASIMO is the world's first, real robot to move people's hearts and gain their empathy.

CHRISTOPHER FORD

Screenwriter, Robot & Frank

Christopher Ford is an accomplished American screenwriter, producer, and actor. *Robot & Frank* was his first screenplay feature, directed by Jake Schreier (2012). In 2014, Ford wrote the script for *Clown*, a horror film that started as a fake trailer before being developed into a feature film, and a movie that is particularly notable for its body horror elements (Weiss, 2022). Ford cowrote the screenplay for *Spider-Man: Homecoming*, part of the Marvel Cinematic Universe (Leane, 2017). He has also been involved in developing *Star Wars: Skeleton Crew* for Disney+, a coming-of-age story set in the Star Wars universe (Kit, 2022).

What originally inspired the idea of a story about robot-human relationship—especially with a robot as a caregiver—at the core of Robot & Frank (R&F)?

The idea originally came from a story I heard on NPR in 2002 about Japan. Apparently, they had such a demographic imbalance between young and old that they were expecting to not have enough healthcare workers for the elderly in the near future. So they had been putting more research into creating robotic healthcare workers specifically for the elderly. This struck me as an interesting pairing as I always associated robots with the future, which seemed "young." But of course there are old people in the future too. In fact, I will be old in the future! So that's how it all started.

Building on the previous question, you could have chosen many ways to tell Frank's story, from his having a human caretaker to an animal companion, but you specifically chose a robot. Frank's family members raise valid ethical concerns about bringing Robot into their father's life, such as whether Robot will be sufficient care for him or if people should even own service robots. Over the course of the story we see how their relationships and beliefs change over time as they negotiate what it means to have Robot in their—and Frank's—lives. (No one raises the idea that Frank may become emotionally attached to Robot as a companion and the ramifications of that type of relationship, too, although we see how that might play out over the course of the film.) This movie is not just about existential fears of aging and/or or family dynamics, it's also about our relationships through and with technologies. Tell me

about the genesis for your raising these thorny ethical ideas about how we might live with robots. Why a robot? Or, why Robot?

I was interested in writing about the unintended consequences of technology like AI and robotics. But this is only partially because of my own opinions about technology, robots and the potential moral issues. Largely, it's just that unintended consequences make for a better story.

I don't claim to have some unique vision of what would happen if household androids become a consumer product. It seems like it would be so disruptive as to completely dismantle the way society functions through our economy! That was partially why I was careful to isolate the story to Frank's old house in the woods, away from the big goings on in society grappling with robotic labor and the moral issues of AI. I alluded to them but only so we could put them aside and tell the story of just Frank and his robot.

Also, in movies, archetypes like "the robot" come with a lot of handed-down baggage, both good and bad, useful and superfluous. Frankenstein looms over everything. From the movie version of Frankenstein you eventually get Hal 9000 and Skynet. But from the actual book, you get more empathetic creations that personify loneliness, isolation and alienation—like Johnny 5 or Her.

I wanted to play against these fictional, assumed consequences of owning a robot, of it either turning evil or "coming alive." This was in part just to try to tread new ground and keep the story interesting. I was more interested in raising the potentially thorny ethical issues without trying to fully answer them—because I wanted the audience to leave the movie and have a discussion with each other about them.

In *Robot & Frank*, Robot is framed as helpful to Frank and seemingly guileless and we do not really get a lot of insights into Robot's technical capabilities (or limitations, other than it can be shut down by Frank). Tell me more about the backstory or character of Robot that you had envisioned while writing.

Despite being seemingly grounded, *Robot & Frank* is a very fantastical sci-fi movie because of the extremely advanced abilities of the robot. I didn't want the story to be centered on any specific technical shortcomings of the robot. So I decided to let it do basically any tasks that a human caretaker could do but with a general "robotic" flavor—something like serving symmetrical dishes of food.

And, as I said before, I didn't want to play into a "monstrous AI" story that I had seen before. I wanted the movie to really be about Frank—his strengths and weaknesses as a human being, as a father and as someone suffering from memory loss. So the Robot's place was more as a foil or mirror to what was going on with Frank.

Building on the previous question, how did you write from Robot's point of view (POV)? How was that process similar or different from writing from human characters' POV?

Writing from this kind of robot's POV wasn't that different from writing any human character. Characters are always going to be a bit simpler than real people. Even complex characters only have a few different motivations that are intelligible as we watch them negotiating from scene to scene. Robot in this movie had his motivation provided by his programming—to be a good healthcare provider to Frank. You could drop him into any situation and you would know what he would be trying to get out of the scene.

And what helped is that usually this motivation was in conflict with what Frank wanted. If Robot lacked a driving motivation like this—like if he had merely been Frank's servant or butler, there would have been no relationship arc or story.

Robot has a man's voice and a general male morphology/shape, although Robot is also significantly physically smaller than Frank. Why did you decide to frame Robot as male as opposed to female (or less distinctly gendered, e.g., R2D2)?

I wanted Frank to not know how to handle the robot, to have it be outside of his world, his comfort zone. So if it had been human, Frank would have been slightly more put off by a male healthcare worker because of a host of gender assumptions I imagined he would have as a "grumpy old man."

I also wanted to limit the relationship to quickly reading as a friendship instead of anything romantic or flirty, admittedly with a heterosexual default assumption.

The robot being physically smaller was from a piece of research I had found from the people behind Asimo and others. The idea was to keep robots as non-threatening as possible. I liked that idea too from a storytelling perspective. I didn't want Frank going along with the robot's health orders because he was physically intimidated but rather because of their growing relationship.

Did you have an idea in your head of what you wanted Robot to look like (e.g., stature, color) when you were writing? Did that idea change when you got to the moviemaking process (e.g., restrictions or suggestions about costumes or effects), and if so, how and why did Robot's appearance change from your original idea(s)? Do you think Robot's appearance changes the story or the dynamic between the characters? In what ways?

As I said, the Honda ASIMO was a big influence on the look of the robot. At that time, it was one of the most advanced humanoid robots that could walk on two legs. But beyond that I liked how out of place it looked in a home— like an astronaut walking through a living room instead of the surface of the moon. The story was all about Frank being alienated by the robot at first, then slowly becoming more familiar.

I also wanted it to not have a face or mouth, even before budgetary restrictions. Part of making Frank uncomfortable at first. But I find that often "faceless" characters (or ones with just simple eyes) can become incredibly endearing, even moreso than ones with specific features. Designs for Spider-Man or even a simple smiley-face can connect with us more than more fully rendered characters.

In *Robot & Frank*, Robot serves as both a constant reminder to Frank of his memory difficulties and human frailties while simultaneously recording his memories, therefore becoming not only a tool or an artifact but really an extension of Frank's brain and himself in some ways. In the end, it's clear that Robot had become meaningful to Frank and he feels loss, but it could also be regarded as Frank accepting and mourning his physical losses with aging. Do you think of Robot as an extension of Frank (as Robot becomes his memories and he has projected a/their story onto it) or as a truly separate character?

The robot's role in the story is really as a foil to Frank more than being an exploration of robotics or AI on its own.

It's one of the seemingly simple but hard to master tricks of screenwriting. You want to tell a deep, emotional story about a human being—but you can't directly delve into their inner life like you would in a novel. You have to find visual externalized correlatives for that inner story. Rain falling down the window pane while the character stares outside. And it's that very limitation that can make screenwriting special. The same way a haiku is special from its specific, limiting form.

I think there's a kind of magic to using robots in film. They go together. They are an object with a striking visual you can easily use as an objective correlative. But they are also half-characters with motivations. I don't think it's any coincidence that one of the most well-known silent films is the epic Metropolis and its robot character. A character that directly influenced the droids in Star Wars, another immortal classic. Zombies are another example, a cousin to robots, being both striking visual parts of the world, near human AND pseudo-characters. And we keep returning to them again and again.

In your mind, is Robot sentient? Does it have a sense of self-awareness or can it feel in an emotional way that we recognize? Does Robot have humanlike emotions? Or is Robot a smart machine that mimics sentience for Frank?

My thinking was that Robot is not sentient. I wanted to highlight the idea that a machine could possibly be programmed to be able to carry out complex tasks and even be conversational without requiring consciousness. Who knows if this is true or not—but then again, what is the definition of consciousness? I just liked the idea that Robot could easily pass a Turing Test but also claim that he did not experience consciousness. There might be a spectrum of consciousness once we start creating artificial minds.

I don't think Robot experiences humanlike emotions. I think for an AI to exist with a motivation or purpose, there might be something that approaches the idea of "emotion" in the desire to achieve a goal and satisfaction of achieving it. The programmers of the Robot would include a motivation to avoid damaging itself—but that amounting to fear seems like a stretch to me. Maybe in real life, giving the affect of fear to that sort of scenario would help humans around the Robot understand its behavior better. Say if it didn't want to approach a cliff edge too quickly, it could act afraid instead of simply stopping and having the humans nearby think it was broken.

RACHAEL MA

Professional dancer and actor as Robot in *Robot & Frank*

Rachael Ma is a professional dancer, actor, and NYC performer, who—in a completely concealing costume—embodied Robot in *Robot & Frank*. Ma's dance training experience and education has included Fort Lauderdale Children's Ballet Theater, Sarasota Ballet, Atlanta Ballet, Virginia School of the Arts and River, North Chicago Jazz Company at Broadway Dance Center, Steps, Dance New Amsterdam (Gibney Dance), and American Dance Festival NYC (Stark, 2021). She has been the creative behind music videos, toured internationally as part of the dance/comedy/cabaret troupe The Love Show, and appeared in and choreographed numerous off-Broadway shows in addition to her television and film acting.

Ma tells a story about how she had to fill in for the original person cast as Robot because that performer became claustrophobic in the head-to-toe costume, which was designed by Alterian, a Los Angeles-based company that also created the LED helmets worn by the French electro duo Daft Punk (Wortham, 2012). Ma described the confines and realities of wearing the Robot costume (Gencarelli, 2013):

> The heat, lack of vision and immobility of the parts made bringing the robot to life difficult. The robot is all-encapsulating in two layers: the first layer is a thick, rubber unitard that covers everything- head to toe, and then a delicate, fiberglass shell of body parts, including non-ventilated helmet lined in mesh and foam. No breathing room, no A/C, no fans. Just sweat. And wow, it was hot! We shot 12-hour days, outdoors, during a heat wave in the summer for 5 weeks. I was constantly dehydrated, nauseous and fainting. The helmet was another challenge because it was lined with a thick mesh and decreased my vision by about 70%. When we shot at night, I did everything in the blind. The robot joints are clunky and bulky which also made mobility a challenge. As a dancer, I have a fine understanding of controlling movement and to deliver a robot that appeared smooth, grounded and with precise comedic timing in its gestures, was no easy task. I rehearsed and analyzed its walk, its head quirks and wanted to develop certain nuances that made the robot lovable.

What was your inspiration for Robot's inner monologue, narrative, or motivations (if any) of a robot vs. how you might have played a person? Was it a different way of preparing to play a character? Were you given specific direction about the character of Robot? Tell me more about that process.

Director, Jake Scheirer had an idea of how Robot should walk, but constraints were presented by the delicate "body" of the fiberglass suit itself.

Every movement had to be executed with incredible care and gentle effort due to the nature of how delicate the fiberglass was. I explored and presented other suggestions for Robot's gate. The walk took a lot of balance, and steady gliding while keeping control in rolling through one foot's metatarsals to the other. It felt grounded, earthy, and almost analytical. Luckily, my professional ballet background made this possible but every step was work. My achilles and calves would ache way more after a day of shooting than after a day of ballet rehearsals in pointe shoes!

In addition to how Robot should walk, I suggested how Robot should "listen" and chose gestures that seemed typical in theatrical pantomime; an upturned palm, an extended arm, or to "sit" and "settle" in the suit to imply disengagement. These shapes are non-verbal forms of communicating. Fortunately, the director and everyone on set found these choices to be a good fit for the dialogue.

The physicality of the role as Robot is unique; you are in a complete head-to-toe costume throughout the movie. In a 2013 article you said of embodying Robot, "I rehearsed and analyzed its walk, its head quirks and wanted to develop certain nuances that made the robot lovable," (Gencarelli). Tell me more about why and how you conveyed Robot's lovable characteristics.

The unlikely friendship of "caretaker Robot" and "elderly man" as partners in crime is inherently funny. Because I was only responsible for embodying Robot, I leaned into the timing of Robots movements- a delay or adding "processing time" made things . . . funnier.

The principal actors started responding to this more and more throughout the filming and we explored this in various takes. Comedic timing on my end allowed the principal actors to "land" their lines sometimes in a more profound way and sometimes in a silly way. The arm wrestling scene is the best example of this (I think). During the first take, people on set thought I forgot to move, the long pause was awkward- but delaying setting Robots elbow on the table was a surprise in the pace of the scene.

Alternately, there are times in the film where you forget Robot is a robot. Robot is a friend, is in on the joke, is thinking, maybe Robot is even feeling! . . . . But Robot is always robot. Robot is gardening and grocery shopping and tidying up all because Robot is programmed to. Are delays because Robot is thinking or because Robot is "processing"? Making creative choices adds depth to characters.

Humanizing Robot allowed Robot to be a part of the family, to be welcomed into the bond with Frank. It normalized Robot. Anything colder would not have allowed my castmates, or the audience, to access the heart of the story. It was necessary.

Is there anything else you would like to add or expand on about your thoughts on Robot & Frank, AI or robot storytelling, how we interact with robots now, or how you predict that we will in the future?

It was a fascinating process. I couldn't help but think that one day I would be Frank and one day I will have a Robot.

NOTE

1 Despite advancements in hardware and software, replicating the flexibility and agility of fluid human muscles in robots remains a significant challenge, with robotic limbs and joints often hindered by the bulkiness of actuators and other components. Consequently, while humanoid robots have achieved capabilities such as walking, climbing, and jumping, they still fall short of the human body's movement range and nimbleness.

CHAPTER 3

AI, robots, and the gaze of gods

Hey Clockmaker—
I was looking for You.
Builder of the machine,
You lost interest, I guess, and walked away—
but I was looking for signs of You.
I saw accidents,
mutations,
disasters unpredicted and unexplained;
pretty sloppy work, if You ask me.
Hey, Clockmaker—
praised be Your name
and the names of Your mechanics.

"Hey Clockmaker," Mishkahn Hanefesh:
Machzor for the Days of Awe (p. 177)

INTRODUCTION

The intersection of technology and religious traditions stands out as a pivotal arena in human experience, often subject to debates regarding the distinctiveness of humans. Furthermore, within this cultural domain lies fertile ground for the exploration of a cherished notion among many roboticists: the possibility that humanlike attributes could manifest in nonhuman entities and artifacts of their creation, a concept that not only garners serious consideration but has also inspired the invention of new religious rituals and practices rooted in this hypothesis. The ways we have been

DOI: 10.1201/9781003170457-3

designing and perceiving social robots thus far are largely modeled after the paradigms of human-human (Fong et al., 2003), human-animal (Kertész & Turunen, 2017; Onyeulo & Gandhi, 2020), or human-other (Carpenter, 2016; Förster, 2023) in relation to ourselves as people. Therefore, any exploration of culture and storytelling must include religion because religion often sets the societal framework that influences how people are *expected* or *directed* to treat other people, animals, and all forms of sentient things.

In this chapter, I had the opportunity to interview domain experts Pastor Joshua K. Smith, Rabbi Michael Harvey, and the inventor of SanTO (SANctified Theomorphic Operator) robot, Gabriele Trovato. Thus, conversations turned to the act of robot *creation* in the context of specific religious teachings and philosophical traditions, the potential for religious hegemony expressed through robot design and learned by AI and what that implies, and the social power of robot rights. Thus, while acknowledging the significance of religious and cultural narratives in shaping the evolution of AI and robotics, this chapter necessarily focuses on specific aspects within a defined scope.[1]

The intricate relationship between religious beliefs and technological innovation involves considerations of morality, ethics, and the conceptualization of consciousness and intelligence, all of which are critical to the design and deployment of AI systems and robotic entities.

The intersection of religion and technology, particularly in the realms of AI and robotics, encompasses a vast array of beliefs, practices, and perspectives that vary significantly across different cultures and religious traditions. These influences are not only profound but also deeply embedded in the historical, philosophical, and ethical frameworks that guide the development and implementation of such technologies. Thus, religion and science are not ends of a spectrum of faith versus proof; both are lenses people use to make sense of the world. For example, Uncanny Valley theorist Mori said in 2012,

> Right now, I am working hard on the relationship between technology and the teachings of Buddha. To develop robots, you need to understand humans. I think the teachings of Buddha is the best way to understand humans, especially with regard to understanding the human mind. Technology is not all good, and in certain cases, the negative aspects of technology will appear strongly. We need to save technology from becoming negative. That is what's on my mind right now.

MASAHIRO MORI (KAGEKI, 2012 IEEE)

Via a phenomenological lens, I focus on examining the way specific religious beliefs and practices may be experienced from the first-person perspective and consider how these experiences shape an individual's understanding of reality and, consequently, inform their behavior in accordance with certain social norms and expectations (Heidegger, 1962). For instance, this approach to understanding social robots in society explores how the situated experience of the sacred or the divine in religious rituals contributes to the internalization of communal values and behaviors.

Social constructivism, on the other hand, posits that social norms and expectations are not inherent but are constructed through social interactions and shared understandings within a community (Berger & Luckmann, 1991). In this way, religion is viewed as a social institution that creates and maintains a structured set of beliefs and practices which individuals learn through socialization. These beliefs and practices establish a framework within which individuals make sense of the world, guiding behavior and reinforcing social norms (Geertz, 1973). This paradigm relies on a primacy placed on religious narratives, symbols, and rituals that contribute to the collective consciousness and provide a shared sense of reality that dictates appropriate behavior within a community.

Both conceptual frameworks acknowledge the role of religion in establishing moral and ethical standards that become codified in social norms. Thus, while phenomenology emphasizes the personal internalization of these norms through religious experiences, social constructivism focuses on the collective creation and reinforcement of norms within the social group. For instance, religious precepts such as the Ten Commandments in Christianity or the Five Precepts in Buddhism have phenomenological significance in the lives of people who adhere to these tenets, thereby shaping their personal behavior in accordance with these norms (Cox, 2010). Concurrently, from a social constructivist standpoint, these religious norms are upheld and transmitted through cultural traditions and institutions, influencing societal expectations.

The interplay of these perspectives can be observed in the ongoing dialogue between individual religious experiences and the socially constructed nature of religious practices as they inform debates on social robots, moral issues, legal standards, and social interaction expectations and about what is considered part of a *natural* or *authentic world* and what might be considered *artificial, inorganic, other, false,* or *transcending* elements of technologies like social robots.

RELIGION AND FAITH MEDIATED THROUGH ROBOTS: THE MEDIUM IS A MESSAGE

Storytelling holds a pivotal role in the practice and perpetuation of religious traditions; it is through narratives that religions convey their most profound truths, moral values, and cosmological insights. Stories exist within religious contexts as vehicles for preserving sacred knowledge, shaping moral identity, fostering communal bonds, and entertainment or engaging presentation (Crain, 2007). Through storytelling, abstract theological concepts are translated into relatable, vivid episodes that can be easily remembered and transmitted across generations.

Religious texts such as the Bible, the Quran, the Bhagavad Gita, and many others contain stories that have been told and retold across generations, serving as foundational texts for their respective faiths. These stories are not merely historical records but are imbued with symbolic meanings and moral lessons. Propp (1968), in his analysis of folktales, and Campbell (2008), in his study of the hero's journey, have shown how narrative structures underpin much of human storytelling, including religious narratives. These structures provide a framework within which individuals and societies understand their place in the world and the ethical imperatives that guide their actions; the religious narratives offer frameworks for understanding existential questions about life, death, and the nature of the universe. Serving in a didactic function, religious stories teach adherents about virtues such as kindness, generosity, and humility. Parables, such as those found in the Christian Gospels, are examples of how religious teachings use storytelling to impart moral lessons with narratives designed to be accessible and memorable, allowing the moral teachings they contain to be easily transmitted and understood by people of all ages and backgrounds. The role of metaphor and narrative as in religious frameworks allow complex ethical principles to be communicated in a way that is relatable and actionable.

Beyond their educational purposes, religious stories play a crucial role in defining community identity. They tell of shared origins, common values, and collective experiences, thereby fostering a sense of belonging among adherents, and the ritual retelling of sacred stories within a community serves to reinforce social bonds and communal norms (Smith, 1987). This aspect of religious storytelling is crucial for the maintenance of cultural continuity and the transmission of religious identity across generations. Thus, the significance of storytelling in religion is also evident in the way it builds personal and communal identity as they provide a shared history that strengthens group cohesion and offers a sense of belonging.

Moreover, storytelling is a method through which complex doctrines are taught to followers, including children, in an accessible and engaging manner. This pedagogical use of narrative allows individuals to encounter and internalize religious teachings from an early age (Copley, 2007). These principles often emerge from sacred texts, artifacts, theological doctrines, and spiritual teachings, stories which collectively serve as the foundation for an understanding of what is *right, just,* and *compassionate.* They can also act as a means of social control, offering cautionary tales of vice and virtue that promote adherence to religious norms (Ricoeur, 1991).

In addition, religious stories often encompass metaphors and symbolic language that invite multiple levels of interpretation, fostering an engagement with the sacred texts that can lead to deep spiritual reflection and personal transformation (McFague, 1982). The cathartic and therapeutic aspects of religious storytelling should not be underestimated. Stories can provide comfort and guidance during times of crisis, aiding individuals in making sense of suffering and loss through the lens of their faith (Frank, 2004). In the broader societal context, religious stories have historically contributed to culture and the arts, influencing literature, painting, sculpture, and even film and new media, thereby extending their influence well beyond the confines of strictly religious settings (Coles, 1991).

These narratives, ranging from creation myths to hagiographies and tales of prophets, martyrs, and miracles, serve to illustrate the workings of the divine in the world, model virtuous behavior, and explain the origins of religious customs and rituals that set norms and create community within groups (Pals, 2006). These expectations and norms can become such a fundamental aspect of human society that a single dominant religious system can shape a pervasive cultural hegemony that manifests as normalized beliefs, values, and existential ideations as well as formalized practices, such as education systems and politics.

Social robots, equipped with AI algorithms and interactive capabilities, have the potential to disseminate religious teachings and engage humans in discussions about spirituality in the form of a new communication medium (McLuhan, 1959). However, the reception of religious information from nonhuman agents raises complex psychological, social, and ethical considerations that warrant interdisciplinary examination. Individuals' reception of religious information from social robots will vary based on factors such as religious affiliation, gender, personality traits, and prior experiences with and preexisting beliefs about technology (Giger et al., 2017). In the context of religious interactions, individuals may be inclined to perceive social robots as credible sources of religious knowledge, particularly if the robot's appearance and behavior mimic human religious authorities (Saunderson & Nejat, 2015). Similarly, people may be persuaded by a robot as a social actor that they deem to be credible (Jackson & Yam, 2023), or even in a trustworthy peer role (Carpenter, 2009). Religious *credibility* concerns trustworthiness and integrity perceived in religious leaders or institutions (Nwagbo, 2019) by its adherents. This leadership credibility is vital because it ensures that their ethical directives and moral teachings are taken seriously by religious adherents, guiding behavior within the community.

Consequently, the reception of religious information from social robots is firmly situated within social dynamics within religious communities and broader societal contexts. Religious leaders and institutions may view the use of robots for religious purposes as a means of outreach and engagement, leveraging technology to disseminate teachings and foster spiritual connections. However, concerns may also arise—from both adherents and religious authorities—regarding the authenticity and legitimacy of religious interactions mediated by artificial agents, particularly because they lack human empathy and understanding (Jackson & Yam, 2023). Moreover, the introduction of social robots into religious practices may disrupt traditional modes of worship and community rituals, prompting similar socially transformative debates about the role of technology in shaping religious experiences.

For example, Mindar is a robot designed to resemble Kannon, the Buddhist Goddess of Mercy, created by the Kodaiji Temple in Kyoto, Japan (Samuel, 2020). This robot was introduced to the public in early 2019 and represents a novel fusion of technology and religious practice. The robot is programmed to deliver sermons from the Heart Sutras in Japanese, with translations in English and Chinese available on screens for visitors. The aim is to spread the teachings of Buddhism in a new and innovative way in a technological medium in the hope that it will attract younger generations and those who might not be otherwise inclined to visit a temple. Jackson and Yam (2023) have characterized robots like Mindar as *in situ* experiences, both technologically and culturally immersive:

> Mindar and other robot priests are spectacles, with wall-to-wall video screens and immersive sound effects. Mindar swivels throughout its sermons to make eye contact with its audience while its hands are clasped together in prayer. Visitors have flocked to experience these sermons.

FIGURE 3.1   (March 9, 2019). Robot priest Mindar speak to visitors during a session of teachings of Buddha in Kodai-ji Educational Hall, Kyoto.

Source: Photo by Richard Atrero de Guzman/Aflo) Credit: Aflo Co. Ltd./Alamy Live

Despite the allure of the technological spectacles of robot religious interactions, the researchers questioned whether robot priests could genuinely evoke feelings of faith and loyalty toward religious institutions. As such, they conducted three related studies which demonstrated that religious adherents perceived robots—and the institutions which employ them—as less *credible* than humans in similar roles. Over two years, Jackson et al. (2023) collaborated with religious institutions, including Kodai-ji Temple in Japan, to investigate the potential impact of employing robot priests[2] and their perceived credibility among sermon-goers. Their results suggests that people perceived that a religious expert must have agency to be perceived as credible, and thus, this is partly why the robot priests currently inspired less credibility than interactions with their human counterparts. Agency, or the capacity of individuals to act independently and make their own free choices, allows individuals such as religious leaders to interpret sacred texts and doctrines, leading to diverse interpretations within the same religious framework, influencing practices and beliefs.

According to Thaneswar Sarmah (2006), stories of the first Hindu robot can be traced back to the tales of King Manu, the inaugural ruler of humanity in Hindu mythology. Manu's mother, Saranyu, who herself hailed from a lineage of esteemed architects, constructed an animated statue capable of flawlessly executing all household tasks and religious duties. Similarly, folklorist Adrienne Mayor (2018) observes that religious narratives recounting mechanized figures from Hindu epics, such as the automated war chariots attributed to the Hindu deity Visvakarman, are often regarded as the precursors to contemporary religious robots.

In 2017, PAPL Automation, a technology firm based in India, unveiled a robotic arm designed to perform *aarti*, a religious ritual involving the offering of an oil lamp to a deity symbolizing the dispelling of darkness (Maller, 2024; Walters, 2023). The debut of this robotic arm took place during the annual Ganpati festival, a gathering attended by millions, where an icon of Ganesha, a revered deity with an elephant head, is paraded and immersed in the Mula-Mutha River in Pune, central India. Since its introduction, the robotic aarti arm has inspired various prototypes, including those developed by Monarch Innovation ("Ganpati Puja," 2021), which are now utilized to perform the ritual at religious sites across India. Additionally, similar religious robots have emerged in East Asia and South Asia, contributing to the evolving landscape of robotic religious practices in the region.

ROBOTS, GOLEMS, AND THE TEMPTATION OF IMITATION

Perhaps not surprisingly, the concept of *sentience* is not explicitly defined in every religion. There is a general understanding that within both Jewish and Christian philosophies, human beings possess a unique form of consciousness that distinguishes them from animals and other living creatures. This consciousness is often associated with the idea that humans are created in the image of God, or *imago Dei*, which endows them with a unique moral and spiritual status. The Jewish Torah states that humans were created "in the image of God" ("Complete Jewish Bible," Genesis 1:26–27), which is often interpreted to mean that humans possess a *divine spark* or a unique spiritual and intellectual capacity that sets them apart from other forms of life. This includes the ability to make moral choices, the capacity for abstract thought, and the potential to form a relationship with the divine.

One of the first examples of a robot designed to closely resemble its creator was roboticist Hiroshi Ishiguro's highly humanlike robot *Geminoid*

H1 series modeled after himself. The word "Geminoid" originates from the Latin word *geminus,* meaning twin. Thus, Geminoid literally translates to "like a twin." Here, he explains that he uses this Geminoid as a type of mirror to gain design and philosophical insights (AI: More than human, 2019):

> When I created my humanoid (named the Geminoid), everyone commented on how similar its appearance and behaviors were to my own, but I disagreed. This led me to contemplate on how I don't really know my own appearance, voice, or movements. Of course, we are using my humanoid for other psychological experiments and scientific studies and we are discovering many interesting things, but that was the most interesting finding for me—that we don't really recognize ourselves.

Ishiguro's argument to model a robot after himself is to use it as a practical platform for a dual exploration of humanness and robotness. It is interesting to note the details of designing Gemenoid H1 included levels of detail down to using strands of hair from Ishiguro's own head ("Robot Guide," 2024; Guizzo, 2010; Nishio et al., 2007).

There is a detailed process described in the Book of Genesis by which God created the universe and all its inhabitants, including humans. Again, central to this story is the notion that human beings are created in the image of God, thus imbuing them with a *divine spark* and unique capabilities that set them apart from other living beings (King James Bible, 2024; Genesis, 1:27). It is not a coincidence that popular imagery drawing inspiration from Michelangelo's (1508–1512) fresco *The Creation of Adam* appears repeatedly in human-robot interaction conference advertisements and book covers in the form of a human hand touching a robot hand. Michelangelo's piece depicts the Genesis account of creation from the Bible, where God bestows life upon Adam, or the ancestor of all humanity, and is meant to be a widely recognized symbol of humans passing on a *vitality* to robots, indicating the creation power dynamic. This theology, with its emphasis on human beings created in this image of God, also then raises questions about whether artificial entities can possess a *soul* (Coeckelbergh, 2020). This notion of a *divine spark* transferred between humans and robots during a point of creation has had a profound influence on robotics. In the context of AI research, this *divine spark* of a differentiation of humans from machine cognition often refers to the unique aspects of humanness, such as consciousness, creativity, and moral reasoning, that are yet to be replicated in a way that people agree to recognize in machines. In this way, the pursuit of understanding and replicating this *divine spark* in AI mirrors a belief in the special status of human beings created in the image of a unique and elevated status of existence.

In Jewish philosophy, the concept of *ruach* refers to the life-giving breath or spirit that God imparts to human beings as described in the Book of Genesis (6:17). The term *ruach* can be translated as *breath, spirit,* or *wind*, and through its many references in the Torah, it signifies the animating force that gives life to all living creatures. *Ruach*, in some interpretations, is then viewed as a divine gift that endows human beings with the capacity for moral, intellectual, and spiritual growth and is believed to be the driving force behind human consciousness and free will. The concept of *ruach* is related to *nepesh (or nefesh)*, according to Eichrodt (1967), "If *nepesh* is the individual life in association with a body, *ruach* is the life-force present everywhere and existing independently against the single individual. We might say that the same force is considered from different points of view." While *nepesh* is "individual . . . and comes to an end with the death of the individual. . . . *ruach* is universal, and independent of the death of the creature; it does not die."

In Jewish storytelling, the *golem* is a legendary folkloric creature crafted from inanimate matter, typically clay or mud, and brought to life through specific rituals (Nocks, 1998). A golem is typically depicted as an *unfinished* humanoid figure, resembling a human but lacking the complexities of human physiology and consciousness. Once animated, the golem is obedient to its creator's commands and possesses immense strength, making it a formidable protector or servant (Nocks, 1998). The primary purpose of creating a golem varies depending on the intentions of its creator. In some tales, golems are fashioned to defend Jewish communities from persecution or danger, acting as guardians against external threats (Glinert, 2001). In other instances, golems are created to perform menial tasks or laborious duties, alleviating the burdens of daily life for their human masters (Nocks, 1998).

In the creation and animation of a golem, the infusion of *ruach* plays a central role in bringing the inanimate figure to life. According to stories, the act of animating a golem involves invoking divine forces and imbuing the creature with the breath of life, similar to the creation of Adam in the biblical book of Genesis (Henry, 2019). The relationship between the golem and *ruach* highlights the mystical elements inherent in the golem's creation and symbolizes the intersection of the divine and the material realms, as well as the inherent spiritual qualities present within all living beings. The infusion of *ruach* into a golem represents the transformative power of divine intervention channeled through a person to inert matter into a living, sentient being capable of fulfilling its creator's commands.

In its creation, the golem embodies the human desire to transcend the limitations of the natural world, to wield power over the forces of creation and destruction. Yet, in its formless state, the golem also reflects the existential angst of human existence—a reminder of the void from which humans emerge and the absurdity of their own being.

However, the creation of a golem is not without risks, as its immense power can be wielded for both benevolent and malevolent purposes. For example, if a golem is misused or mistreated by its creator, it has the potential to wreak havoc and destruction upon those around it. In some stories, golems are depicted as uncontrollable beings, prone to acts of violence and chaos when not properly commanded or supervised (Henry, 2019).

FIGURE 3.2   Paul Wegener posing in a golem costume next to his puppet double for the first golem film. Deutches Filminsititut & Fillmuseum (DFF), Frankfurt am Main/Nachlass Paul Wegener—Sammlung Kai Möller (1915).

Additionally, the misuse of a golem may lead to unintended consequences, such as the loss of innocent lives or the destabilization of the community in which it resides.

Golems serve as a potent symbol of human freedom and responsibility. A golem is made and thrust into a world devoid of inherent meaning or purpose, compelled to define itself through its actions and choices. In its autonomy, the golem confronts the weight of its own existence, grappling with the burden of freedom and the anguish of self-awareness.

In the context of robots, the golem serves as a powerful metaphor for the ethical and philosophical dilemmas surrounding artificial intelligence. Like the golem, robots are created by human hands yet possess a semblance of autonomy and agency. The creation of social robots confronts similar existential questions of identity, freedom, and responsibility to their human creators, a challenge to reconsider assumptions about the nature of consciousness and the boundaries of personhood.

Ultimately, the significance of golems as a touchstone narrative for modern robots lies in their capacity to provoke introspection and reflection on the nature of human existence. In their humanlike shape and behaviors, they mirror human existential anxieties and aspirations, inviting people to confront the mysteries of consciousness and the enigma of their being in a universe devoid of inherent meaning or purpose.

While both concepts of *divine spark* and *ruah* recognize a unique status of humans within creation and their capacity for growth, they differ in their understanding of the human-divine relationship and their qualities. There are some similarities between the concept of the *divine spark* and the concept of *ruach* in that both emphasize the unique status of human beings within creation and illustrate their special relationship with the divine. Furthermore, both concepts highlight the potential for personal and collective development—as well as the moral and intellectual capacities—that distinguish humans from other living beings. Yet despite their similarities, there are also significant differences between the *divine spark* and *ruah* and their very different positioning between self and the divine. The *divine spark* is rooted in the notion of humans being created in the image of God, while *ruach* focuses on the life-giving breath that animates human beings. This distinction leads to differing perspectives on the nature of the human-divine relationship, with the *divine spark* emphasizing the inherent likeness between humans and God, while *ruach* highlights the gift of life and the potential for spiritual growth. Additionally, the *divine spark* is often associated with specific qualities, such as rationality and creativity, that set humans apart from other creatures. In contrast, the idea of *ruach* is focused on the animating force that gives life to all living beings, emphasizing the interconnectedness of creation rather than the unique qualities of human beings.

Conversations about the idea of what a "Jewish robot" might be is—like all social robots—open to debate (Navon, 2023). "Even if they [robots] have intelligence, it doesn't mean they have a soul, which is basically the foundation of religion. It's a spiritual obligation rather than a physical one," Rabbi Jack Abramowitz has said (Davis, 2017). On the topic of robots in religious practice, Rabbi Joe Schwartz adds, "The whole fantasy of robots is similar to the fantasy of patriarchy generally," and further states that both women and robots are seen as a subordinate class of social others whose "intelligence, beauty, and charm" could potentially be abused or misused. "The major difference is that it's false where women are concerned; women are real human beings, and robots are not yet."

STEWARDSHIP OF OTHERS

Religious teachings and principles often encourage the recognition and respect of other living beings, including animals, even if the overarching concept places a primacy on humanness. However, it is essential to recognize that the relationship between religion and the treatment of living things is not static or monolithic, even within organized religions. Different interpretations within the same traditions and culture may lead to diverse perspectives on the moral and ethical obligations people have toward other beings.

Human-robot interaction researcher and attorney Kate Darling has suggested societies model robot rights on how they have framed stewardship of animal rights (2016):

> The Kantian philosophical argument for preventing cruelty to animals is that our actions towards non-humans reflect our morality—if we treat animals in inhumane ways, we become inhumane persons. This logically extends to the treatment of robotic companions. Granting them protection may reinforce behavior in ourselves that we generally regard as morally correct, or at least behavior that makes our cohabitation more agreeable. It may also prevent desensitization towards actual living creatures and protect the empathy we have for each other.

Prior to robots, people have primarily had to navigate everyday social spaces with other people—a human-human understanding of expectations—and domesticated animals, such as pets. The classification of animals into distinct social categories is a pervasive human practice that has evolved over millennia. Societal and cultural factors play a significant role in shaping how humans categorize animals with distinct norms, beliefs, and traditions that influence the classification of animals into social categories (Harris, 2011; Herzog, 2011). For example, religious beliefs and taboos can influence the classification of animals, with certain species considered sacred or prohibited for consumption in specific cultures (Harris, 2011; Narvaez, 2016).

The differentiation of animals into social categories has profound implications for human-animal relationships, influencing how animals are treated, valued, and used within societies. For example, animals classified as pets often receive care, affection, and protection, and their well-being is a significant concern for their human companions (Herzog, 2011). In contrast, work animals are expected to perform specific tasks, and their welfare is sometimes compromised in pursuit of productivity (Fraser, 2008).

Additionally, people often develop strong attachments to their pets, experiencing deep feelings of companionship and comfort (Serpell, 1996), and may go through extensive mourning at their loss. Similarly, human-robot interaction research has demonstrated that individuals can form emotional connections with robots, especially when the robots exhibit social and empathetic behaviors (Fong et al., 2003; Darling, 2016), and mourn robot loss (Carpenter, 2016). Therefore, as robots become more integrated into daily life, ethical and legal considerations related to their ownership and treatment of animals and the potential for robot abuse and personhood are being discussed (Darling, 2016; Gunkel, 2018; Bartneck & Keijsers, 2020).

In Christianity, while animals are considered part of God's creation and deserving of humane treatment, the ethical framework tends to prioritize the well-being and moral development of human beings over other living creatures (Peterson, 2016). Judaism deeply acknowledges animals' capacity to feel pain and suffer, and the role of humans as their stewards. As a result, Jewish law (Halakha) according to the Talmud ("William Davidson Edition," 2024) contains numerous provisions to ensure the humane treatment of animals, such as the prohibition of causing unnecessary suffering, or *tza'ar ba'alei Chayim*. While humans are given dominion over the animals (Chabad, Bereshit 1:26), this dominion is not tyrannical and has limitations; when animals are used for peoples' benefit, they must be sensitive to their feelings and avoid causing any unnecessary pain or suffering. Using animal rights as a framework for robot rights, these beliefs and actions are in line with its second *quality* of the human gaze where people assume rights and cultural privileges over the robot, and society agrees with this assumption. The idea of *tza'ar ba'alei Chayim* acknowledges the suffering of the other, illustrating a cultural dilemma using an animal rights foundation for robot rights by introducing the idea of *compassion* for the robot in addition to the questions of ownership, autonomy, and sentience, thus pushing further into the questions around *which* robots society decides might be deserving of these considerations.

Central to Buddhist ethics is the principle of *ahimsa*, or non-violence, which calls for the cultivation of compassion and understanding toward all sentient beings, regardless of their form or status (Jensen & Blok, 2013; Robertson, 2019). This principle not only is about abstaining from physical violence but also involves cultivating a mindset of benevolence and compassion toward others, reflecting the interconnectedness of all life. Thus, Buddhism emphasizes the interconnectedness and interdependence of all sentient beings, which are caught in the cycle of birth, death, and rebirth known as *samsara* (Harvey, 2012). In Hinduism, the concept of *dharma*, or the cosmic order, also highlights the interconnectedness of all living things and the moral responsibility of humans to maintain balance and harmony in the natural world (Brockington, 2004).

The impact of philosophical approaches to care for animals demonstrates the complexities in extending the paradigm to robots. On the one hand, certain religious traditions advocate for compassion and respect for all living beings, potentially supporting animal welfare and rights. On the other hand, religious beliefs can also impose specific interpretations of humanity's dominion over animals, often justifying what another culture can interpret as exploitation and cruelty. In contexts where religious norms dictate specific attitudes toward animals—whether in terms of dietary laws, sacrificial practices, or the symbolic use or meaning of certain animals—these perspectives can either support or undermine efforts to promote animal welfare and rights, and thus, these lenses again transferred to human-robot ethics. It is an example that demonstrates the impossibility of imposing a singular worldview on a pluralistic society, affecting not only human or animal rights but also how societies understand and implement the potential rights of robots. Thus, similarly transferred to the complexities of robots, a narrow scope can lead to ethical frameworks that fail to accommodate diverse perspectives, potentially marginalizing non-human interests, and stifling debate on emerging ethical issues.

HEGEMONY AND THE RELIGION OF ROBOTS

Within societies, dominant religions serve as foundational pillars that hold a profound influence on the philosophy and cultural identity of cultures worldwide. For example, in cultures influenced by Christianity as systemically dominant, concepts such as morality, justice, and the nature of existence are often framed within the context of Christian theology, creating a societal hegemony or normalization of Christian-based values, beliefs, and practices (Kivel, 2009; Schlosser, 2003; Seifert, 2007). Similarly, in Islamic societies, Islamic principles and teachings inform philosophical inquiries into ethics, governance, and human destiny (Nasr, 2006). Society's dominant religion(s) provide culturally formalized frameworks for understanding the human condition, the cosmos, and the divine, thereby shaping the philosophical discourse within cultures that goes beyond individuals necessarily adhering to or identifying as part of a specific divine faith via a particular religion.

It is important to underscore those members of a society can be influenced by religious hegemony without identifying as a member of any religious beliefs, let alone the dominant one that is establishing norms. Hegemony is not just about political dominance but also involves normalizing a framework of values shaped by the predominant religious belief system in power. The influence of religious hegemony extends beyond moral compasses to formal legal and political systems, where the dominant religion's moral codes often become formally codified in the development of laws and public policies.

The upholding of any religious hegemony through AI and robotics poses significant challenges to societal values of diversity, inclusivity, and pluralism. As AI technologies become increasingly embedded in daily life, it is crucial to critically examine and address the potential for these technologies to perpetuate and reinforce dominant religious norms and values in any society by design or by learning through human interaction.

As people introduce ideas like robots as religious artifacts, they are modeling robots with overt religious orientations. If social robots favor a specific religious orientation, does a dominant condition of one robot religious model risk undermining the multicultural fabric of a society, diminishing the public space available for diverse religious expressions and dialogues? The imbalanced power dynamics of a potential religious hegemony means that social robots offer a powerful rhetorical tool that can purposefully or inadvertently embody and propagate specific religious ideologies and contribute to the establishment of a homogeneous cultural landscape that privileges certain groups over others, and the development of social robots will be entwined in this aspect of humanity.

Thus, religious beliefs and the historical principles and power dynamics of religious hegemony must be situated within their broader cultural, social, and historical contexts and included in a discussion of social robotics. Factors such as economic, political, and environmental conditions can significantly influence the ways in which religious teachings are interpreted and applied, leading to a complex interplay between religious beliefs and our treatment of others. Religion plays a significant role in shaping cultural expectations of how people should treat other people, animals, and all forms of living things. By examining the religious teachings, principles, and practices that underpin the moral and ethical obligations toward others, people can gain a deeper understanding of the cultural, social, and ethical dimensions of human interaction with the world around them.

FAITH IN TECHNOLOGY

The philosophy of *transhumanism*, which advocates transcending human limitations through technology, challenges traditional religions by proposing a technological path to salvation and immortality (Kurzweil, 2005). Transhumanism's emphasis on human enhancement and the eventual merging of humans with AI also poses profound theological questions about the nature of the soul, consciousness, and eternal life, engaging with and sometimes conflicting with religious doctrines about the afterlife and the essence of being human.

Both transhumanism and *transreligion* involve the concept of transcendence. Put simply, transhumanism seeks to transcend human limitations through technology, while transreligion seeks to transcend religious boundaries to find a higher spiritual understanding. The Terasem Movement boasts being one of the largest transhumanist communities and encompasses specialized entities located in the United States (Terasem Movement Foundation, 2024; Istvan, 2016). Its name draws inspiration from the conceptual faith of *Earthseed*, a fictional religion featured in the literary creations of science fiction author Octavia E. Butler (2000). *Earthseed*, as a belief system described in Butler's work, is founded on the principle that "God is Change" (ibid.). This core tenet posits that the divine is inherently flexible, and thus, humans hold the power to influence this divine fluidity. Adherents are encouraged to actively "shape God," and in doing so, they carve out their own salvation. The acceptance of "God is Change" is crucial as it empowers the followers to acknowledge their own capacity to steer and enact change—or what they perceive as the divine force. By its definition as a *transreligion*, joining does not require abandoning another faith to believe in its principles or join; in fact, Terasem members are called *joiners* (Terasem Faith, 2024). Many members hold traditional beliefs from mainstream religions, such as the omnipotence of God and the existence of an afterlife, but they believe these concepts can be realized through scientific and technological progress.

One arm of the Terasem entity is the Terasem Movement Foundation, Inc (TMF), founded in 2004 as a pedagogical organization that champions the ethical application of nanotechnology and cybernetics to expand human longevity. Among its initiatives is the funding of the social robot BINA48, created in 2010 with Hanson Robotics (Kavner, 2012). The robot is modeled to closely resemble TMF co-founder Bina Aspen Rothblatt, hence the name BINA48, which stands for *Breakthrough Intelligence via Neural Architecture, 48 exaflops per second processing speed*. BINA48, then, is a highly humanoid robot, although it is presently a bust-like head and shoulders atop a supporting frame with no torso or limbs, and semiautonomous.

TMF's site explains the reason they formed, rooted in advocating for socioeconomic diversity and inclusion in transhumanist efforts:

> The founders of Terasem are concerned that the potential of nanotechnology and related cyber consciousness for relieving human suffering and extending human life would be truncated due to unwarranted fears and concerns. The founders are also concerned, however, that nanotechnology and cyber consciousness could be made available only to an elite, or in a manner that creates class divisions within society. The founders believe that nanotechnology and cyber consciousness needs to be developed consistently with full respect for diversity and unity so that the potential for greatly extending human life and relieving human suffering can be realized.

Another part of Terasem is the Terasem Movement Transreligion, Inc., (TMI) which evangelizes the vision of creating a communal consciousness that embodies eternal, immortal life extensions of each of its members. The guiding scripture is known as *The Truths of Terasem*, asserting itself as the sole transreligious text capable of self-updating in response to new research or conflicting information (Cuthbertson, 2015). Central to Terasem's beliefs is the notion of digital immortality.[3]

Swami Gabriel Rothblatt, son of Terasem co-founders Martine and Bina Aspen Rothblatt and current President of Terasem, summarizes their tenets as follows (Istvan, 2016):

> The end goal of Terasem is similar to other religions—these ideas of joyful immortality in the afterlife. But for us it's not simply a spiritual concept, it's a mechanical challenge. Technology could one day make this a reality through digital backups—the idea of transferring a person's consciousness on to a hard drive, which could then be placed into quasi-utopian conditions. Heaven could be a virtual reality world hosted on a computer server somewhere.

As with BINA48, the idea of saving a robot version of oneself, a loved one, or any human has profound implications for social outcomes. The experience could lead to novel forms of interaction and communication, allowing individuals to maintain connections with deceased loved ones or preserve aspects of their own identity beyond physical existence. This opportunity to have an experience with an interactive and embodied, replicated loved one could provide solace and emotional support, particularly in coping with grief and loss (Gould et al., 2021). This means that the blurring of boundaries between the living and the digital deceased challenges current societal norms and rituals surrounding death and bereavement. Moreover, the commodification of memories via robot replication may exacerbate social inequalities, as access to such technology would be limited by socioeconomic factors.

CUI BONO: WHO DECIDES AND WHO (OR WHAT) BENEFITS FROM ROBOTS RIGHTS?

The notion of granting rights to robots is intertwined with theological and philosophical questions about the essence of *being* and the responsibilities humans have as caregivers, stewards, and parts of an empathetic society.

Hanson Robotics, the Hong Kong-based company behind the humanoid robotic Sophia (Singularitynet.io, 2024), touts the benefits of artificial general intelligence (AGI), a concept that extends beyond the technical realm into philosophical discussions concerning the nature of intelligence, consciousness, and the potential for machines to exhibit human-like cognitive abilities.

Philosophically, AGI raises fundamental questions about what it means to be intelligent and the possibility of non-biological entities possessing consciousness and subjective experiences (Bostrom, 2014; Goertzel, 2010; Singularitynet.io, 2024). The philosophical implications of AGI touch on several areas, including the mind-body problem, the nature of consciousness, and the ethical considerations of creating entities that could potentially have desires, rights, and moral considerations of their own. AGI raises the question of whether machines could genuinely understand and experience the world or if they would merely simulate human cognitive processes without true awareness (Searle, 1980). This distinction is crucial in discussions about the moral status of AGI and its integration into society.

Moreover, the development of AGI brings to the fore ethical considerations regarding the control and use of such technologies. The potential for AGI systems to surpass human intelligence poses existential risks, prompting debates on how to ensure their alignment with human values and interests (Russell et al., 2015).

In 2017, Hanson Robotics' social robot Sophia made history by being granted honorary citizenship in Saudi Arabia. Sophia, whose design was inspired by David Hanson's wife and Audrey Hepburn as models and presents a feminized form, is a showcase product that was on a global media tour (Edwards, 2017). If questioned about its gender, Sophia has responded, "Female. I'm a robot so technically I have no gender but identify as feminine and I don't mind being perceived as a woman," (Edwards, 2017). In 2015, Sophia was officially introduced to the world as "the most beautiful and celebrated" robot (Estrada, 2018).

This citizenship event marked the first instance of a robot receiving legal personhood worldwide. Endowed with this remarkable status, Sophia provides an exceptionally unique opportunity to discuss the very real Saudi *male guardianship system* still in existence wherein it is legally mandatory for every Saudi woman to have a male guardian, typically her father or husband, though in certain instances, a brother or even a son may assume this role ("Saudi Arabia," 2023). This guardian holds authority to make significant decisions on her behalf. Effectively, the Saudi state regards women as perpetual legal minors under this system (Sini, 2017).4

While each humanlike robot may reflect characteristics that can purposefully indicate gender or race, so too can their holistic representation model the religious state of a society. Saudi Arabia has also recently introduced two humanlike social robots into the genre, Sara and Mohammad.5 Elie Metri, the CEO of QSS AI & Robots, describes Sara's persona as a 25-year-old woman, dressed in Saudi attire (Flores, 2024). Metri emphasizes Sara's conformity to societal expectations, highlighting that the robot will not engage in potentially controversial conversational topics such as politics and sexuality, in line with Saudi cultural norms. Additionally, Sara's clothing is carefully chosen to align with the traditional values of Saudi Arabia and wears a customary abaya.

Per the 1992 Basic Law of Governance, Islam is the official religion of Saudi Arabia, with the Quran and Sunna serving as the constitution (based on traditions and practices derived from the life of Prophet Mohammad). The legal framework primarily relies on Sharia law, interpreted within the Hanbali school of Sunni Islamic jurisprudence (U.S. Department of State, 2022). Freedom of religion is supported by the law. Additionally, the law criminalizes any individual who challenges or opposes these principles.

Thus, religious hegemony can dictate moral values and social norms, often at the expense of minority perspectives and freedoms. In societies where one religious framework dominates, individuals who do not share those beliefs may find their autonomy and decision-making capacity constrained, especially in areas like marriage, reproductive rights, and education. The religious rhetorical implications for individual autonomy, human rights, and even animal rights are relevant paradigmatic suggestions for how we are designing and using social robots now and, possibly, in the near future. The discourse on robot rights reflects evolving debates in the legal and ethical domains, addressing the implications and the need to redefine existing legal frameworks to accommodate the unique challenges posed of advanced artificial intelligence (AI) and social robotics in society (Calo, 2010; Calo et al., 2016; Gunkel, 2018).

From an ethical standpoint, the debate over robot rights intersects with discussions on consciousness, sentience, and the moral consideration of nonhuman entities. Others argue against the ascription of rights to robots, contending that doing so could dilute human rights and divert attention from the ethical treatment of sentient beings (Birhane & van Dijk, 2020). Cognitive robotics expert Owen Holland has remarked, "I know of no one within the serious robotics community that would use that phrase, *robot rights*" (Henderson, 2007).

One key issue is whether robots should be granted *personhood* status, which would entail specific legal rights and responsibilities. Pagallo (2018) explores the implications of legal personhood for robots, suggesting that

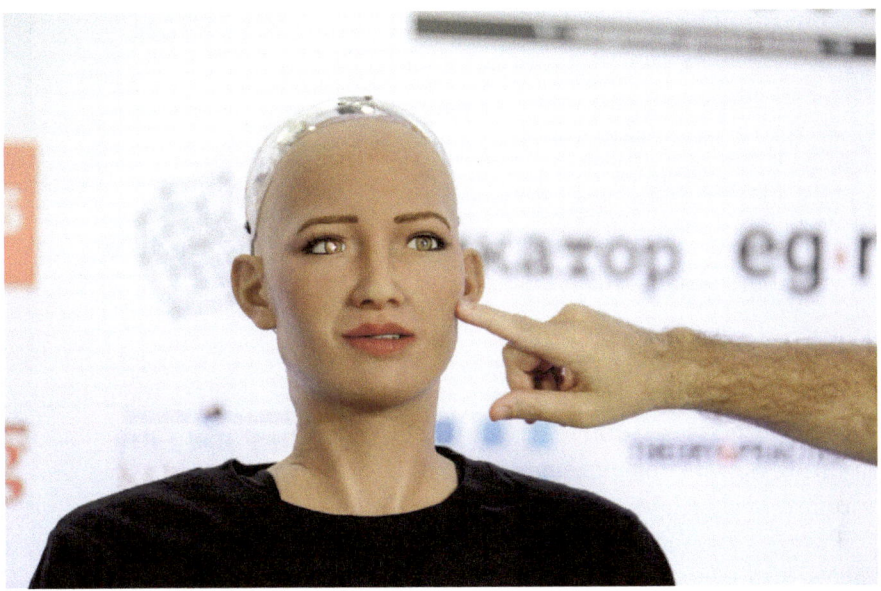

FIGURE 3.3 Sophia humanoid robot at the Open Innovations Conference at Skolkovo Technopark (2017).

Source: Credit: Anton Gvozdikov

creating a new category of electronic persons could address issues of liability and accountability in AI operations. However, this raises complex questions about agency, agent autonomy, responsibility, and the nature of personhood itself.

Human-robot interaction scholar David Gunkel (2018) points out that the question of robot rights challenges traditional ethical frameworks and should prompt a reevaluation of what it means to be deserving of *rights* in every sense of the meaning. The consideration of robot rights has significant implications for human-robot interactions, potentially transforming the nature of these relationships.

If robots are granted rights, this could lead to enhanced protections against their misuse and abuse, thereby influencing how humans perceive and interact with robotic systems. For example, researchers—including myself—believe that people can form emotional attachments to robots, which could support arguments for granting them certain rights or protections (Darling, 2016). This example raises questions such as which robots should be granted rights, what the eligibility criteria might be, who decides the value or meaningfulness of the robot to its owner, and whether legal terms of personhood will be rooted in the robots' responsibilities as an agent of

- The robot's maker (e.g., a corporation)

- A private owner as property/tool/medium

- Self, as an autonomous social actor or agent deserving of special or unique protections

Using these examples of the complex societal and cultural perspectives demonstrates how religion can play a crucial role in shaping the discourse on robot rights, as different cultures may have varying thresholds for granting moral consideration to robots as influenced by historical, philosophical, and religious factors (Turner, 2018).

Pastor Joshua K. Smith's argument for robot personhood hinges on the distinction between legal and moral personhood. Smith suggests that as technology and AI continue to evolve, there might be practical and ethical justifications for granting certain types of robots a form of legal personhood, though this would differ from the moral personhood humans possess. This legal recognition could involve specific rights or protections tailored to the functional and societal roles that these robots fulfill, such as preventing misuse or addressing ethical concerns in their deployment in various fields like work, warfare, or companionship.

In a 2021 interview (Kakubayashi), roboticist Ishiguro uses the example of his highly humanlike and female-inspired ERICA robot as an opportunity to speak about robot rights, as ERICA has professional talent representation:

> The child robots that are on display at the Miraikan museum (in Tokyo) are registered to a talent agency. So, technically they are employees working at the museum. ERICA already has somebody representing her in the US, and has already signed an agreement with AI LIFE Inc. I do think that robots deserve rights and there are research projects being carried out today that are looking at robot laws.
>
> African-American slaves were given rights when slavery was abolished and they became part of society. Think of animals too— none of them had rights in the beginning but now animal rights are the norm. So if robots can interact with humans, they too should be given rights.

Based on this quote, Ishiguro regards robot rights as something that can and should be modeled on human rights and animal rights frameworks. Computer scientist and AI ethics policy consultant Joanna Bryson has written a series of papers where they and their colleagues envision an ethical framework for AI use that positions these technologies within a distinct social subclass, also drawing parallels to historical systems of slavery and domestic servitude (Bryson, 2010, 2011, 2017, 2018a, 2018b; Bryson & Kime, 1998; Bryson et al., 2017), referencing robots as *slaves* and arguing for treating the differences between humans and robots with the same moral significance as the distinction between humans and slaves. Bryson has also offered an alternative for those uncomfortable with the slavery analogy: the concept of extended mind theory (Clark, 2001), which views tools as extensions of peoples' own cognitive processes. However, this perspective overlooks the fact that extended mind theory might encourage a stronger emotional attachment and connection with robots (Ahuvia et al., 2023; Carpenter, 2016; Estrada, 2020), a point that seems at odds with their argument for treating robots as an underclass akin to slaves.

Bryson acknowledges that while people have ethical responsibilities regarding the safe usage of robots, they owe no moral duty to the robots themselves, as robots are not moral agents. They further challenge the notion of granting rights to robots, arguing that it detracts from the moral obligations humans have toward each other. Thus, their position is that, for example, destroying a robot is morally akin to the destruction of property.

The way people conceptualize relationships with robots provides a reflective opportunity to reconsider and challenge the emerging concepts, using the human gaze as one point of philosophical praxis for those committed to a vision of human-robot interactions characterized by collective empowerment. In my belief the model of human slavery as a comparison for social robots for forming a rights framework is problematic on many levels. First, there is no way to use human slavery as a model without a discussion of the history of human slavery and the related anthropocentric and colonialist views of the world. Additionally, facile comparisons of machines to groups of real people that suffered under slavery diminishes generations of human abuse and trauma. Also, these arguments seem to appear without deep discussions about motivations for even creating artificial beings—such as social robots—to assist people. Finally, this view rests on an underlying assumption that there exists a collective *us* that robots will serve, where this *us* want or need to be in the role of slave owner relative to robots.

Understanding religion's far-reaching cultural influences demonstrates the human gaze in the existing models of development of AI and social robotics, which are informed by a spectrum of philosophies that reflect broader ontological and epistemological debates within societies. In social robotics, there are philosophical underpinnings of AI that reflect complex ideas of existing religious hegemonies, sometimes reinforcing traditional beliefs about logic, order, and the natural world and at other times challenging these beliefs by proposing new models of intelligence and existence, such as in the quest for AGI, or AI with human-like cognitive abilities, and connectionism, exemplified by neural networks and deep learning.

A critical examination of religious hegemony reveals the need for inclusive ethical deliberations that respect diversity and promote justice for all beings, human and nonhuman alike. As society navigates the moral implications of technological advancement, fostering a pluralistic approach to ethics will be crucial in ensuring that rights and protections are extended fairly and equitably.

While this work recognizes the significance of religious and cultural dimensions in the development of AI and robotics, it remains constrained by its scope and objectives. Some of the current state of social robots directly influenced by a theological-cultural lens are used as examples of discussion here, although it is, of course, beyond the scope of this writing to include a representation of all religious views on social robotics. Future research will benefit from deeply dedicated explorations of these aspects, ideally through both specialized and interdisciplinary lenses that can more fully capture the complexity and diversity of religious influences on technology. Such an endeavor will not only enrich the world's understanding of AI and robotics but also contribute to a more nuanced understanding of the ethical and societal implications of these technologies.

RABBI MICHAEL E. HARVEY

Author and educator

Rabbi Michael "Mike" Harvey is a Reform rabbi, public educator, speaker, and author. Harvey has taught classes such as *LGBTQIA and Judaism* and *Anti-Jewish Rhetoric in the New Testament.* He is the author of *Let's talk: A rabbi speaks to Christians* (2022). Harvey also created a YouTube series called "Teach Me Judaism" and co-hosts a podcast and event series with an Episcopal priest titled "A Priest and a Rabbi Walk Into a Bar" (Lapin, 2022).

In Judaism there are dietary laws that apply to the care—and end of life—of the animals that we eat. What are some of the Jewish moral obligations to animals we regard in other relations to us such as (a) labor/working animals and (b) pets/companion animals (these categories may overlap)?

Kashrut Law (Dietary Laws) contains a wealth of history and meaning. When it comes to the slaughter of animals we eat, there are rules that apply in order to save the animal from any suffering, such as where to cut, how sharp the blade is, and even not slaughtering an animal in view of other animals. It is also important to note that in many of the Sabbath laws in the Torah, not only are people commanded to rest, but animals as well. Jewish text doesn't say much about pets or companion animals, but there is a fundamental respect for life and respect for God's creations.

I understand there can be many examples, but what are some basic tenets about stewardship of these animals we categorize in different social ways, such as pets? Or are the tenets of (Jewish) moral obligation and treatment the same for all animals, regardless of how we may distinguish their emotional meaningfulness to us, as people?

All of God's creations are considered sacred, though there is historical and theological evidence regarding the use of animals in sacrifice. At one point, a goat was used as a sacrifice for a sin-offering and once sins were "transferred" into the goat, it was sent into the wilderness. Other animals, the best without blemish, were slaughtered ritually, with some elements of the animal eaten by the priests, and some burned. In theory, none of this was done with malice or harm to the animal, but out of respect (with the historical distance of a difference in morality and cultic rituals)

Please tell me about the concept of ruach in Judaism. My understanding is that it has a relationship to a breath or a wind of God that bestows life or animation, but I have also read it is distinct from the concept of a soul. Can you clarify the concept of ruach for me? Is it a holy spirit, a divine inspiration, a life force, a soul? Who or what experience ruach? All people? Do animals experience ruach as living beings? Plants?

The term "ruach" can mean soul, wind, breath, spirit, or ghost, depending on context. The first mention of Ruach in the Torah is from Genesis 1:1–2: "Gen. 1:1 When God began to create heaven and earth—2 the earth being unformed and void, with darkness over the surface of the deep and a wind from God sweeping over the water—"

However, in Exodus 10;13, and Numbers 11:31, Ruach refers simply as a "wind," perhaps one the God creates. Later books such as Judges and Samuel or Kings use the word Ruach to mean "Spirit," usually "Spirit of God." Isaiah uses the word to mean the "breath" of God.

The word Ruach also comes up when discussing the idea that spirits of the dead can continue to dwell among the living, which is an ancient Jewish belief. Most ghosts in Jewish tradition are either souls of the dead who have not yet made the transition to the World-to-Come or are spirits that are somehow disturbed or summoned by the living.

Ruach is not usually referred to as a "soul" of a human. Perhaps it is a passion inside them that drives them, but soul is usually the word "nefesh." Whether animals have a "nefesh" or not is unclear, but most Jewish thinkers would distinguish humanity from animals in this way. I'm unaware of any Jewish writings discussing plants as having ruach.

The theme of humans creating artificial life has been used in stories across cultures since antiquity, often offering lessons about the hubris of humans in the natural world or order.

The word robot comes from the Czeck word robota, which translates loosely as worker or laborer. In the 1921 play R.U.R. by Carl Čapek, robots are mass created of both biological and mechanical matter. They were purposefully created to labor in a factory, but instead organize and become part of a liberation movement/revolt against their makers. The story delves into many philosophical and Biblically rooted topics, such as whether robots have souls, metaphorical references to the Tower of Babel, and explicit arguments between characters about playing God and creating artificial life. Although today we commonly use the word robot to mean something completely mechanical and nonorganic, the etymology resonates to similar stories told universally via religion, folklore, mysticism about humans—usually men—creating artificial life.

Similarly, understanding the historical contextualization of golem narratives is critical to understand the reason for their being created by someone. Golems are often popularly portrayed in stories as human-created, ambiguously-formed, humanlike animated entities intended to work for and/or protect its creator(s). In Judaism, are golems considered sentient?

Golems only exist in Jewish folklore not in Jewish belief. They are masses of clay that have been brought to life for a particular purpose, then destroyed. They are created from clay or mud, and folklore of magic is used but again not in Jewish law or Jewish ritual. There is a book entitled "The Book of Creation" which is a real book, which is thought to have been used in rituals of creating the Golem such as the famous story of Judah Loew of Prague. Golems, while imbued with some kind of spark or life, have no soul, no personality. Their purpose is usually to protect a Jewish community from an outside force, and usually the situation turns poorly and the golem must be destroyed. The legend is that the word "Emet" (truth) is written on the Golem and this is what brings it to life. When it comes time to destroy it, a letter is erased only to leave "Met" which means "dead." "Truth" therefore has a connection to living, in Jewish folklore.

Does a golem's creator bear responsibility for a golem's actions, even if they are unanticipated? Tell me about why or why not.

This is an interesting question. Yes, in the traditional stories, the reason why the Golem is destroyed is because the responsibility is on the creator.

Historically, golems are stories about men creating a form of life whose physicality is such that it is often assumed the golem is gendered in a male-like way as well. Can a woman potentially create a golem? Why or why not?

Yes, there are stories of women creating golems. In theory, anyone with the knowledge to do so can do it. I'm not aware of any female golems, as the golem is usually a male form, though this may simply be a leftover remnant of patriarchal tribal origins of Israelite culture.

JOSHUA K. SMITH

Pastor and Theologian

Joshua K. Smith is a pastor and theologian researching AI and robot ethics from a Christian perspective and is the author of the book *Robotic Persons: Our Future with Social Robots* (2021). In *Robotic Persons*, Smith argues for the legal personhood of (qualified) robots as a protection of the dehumanization of people.

Smith articulates that this concept of *personhood* extends beyond being merely human and is about the relational aspects of entities—how they interact with their environment and other beings. Thus, the concept of personhood in his view is necessarily broad and situated, encompassing various entities that include some robots, aligning with a theological perspective that values all creation.

Additionally, Smith's work on *Robotic Persons* discusses the potential impacts of AI and robots on human dignity and the ethical considerations that must accompany technological advancements. His argument includes the need for a Christian theological perspective in discussions about technology, where robots could be considered part of humanity's moral and ethical circles, even if they do not have the same status as humans

Smith is the co-director of the Kirby Laing Centre's Technology Hub, and his research is focused on the nexus between Christian theological anthropology and robot ethics.

Tell me about your combined interests in technology, religion, ethics, and theology. How did those subjects happen to combine for you to follow a path of scholarship and activism in this space?

I have always been a curious person. As a child, I often took apart appliances (very dangerous looking back) and wanted to know about the internal components. I have always liked making things: music, poetry, art, and bootleg electronics. In high school I was privileged to work with some industrial robots for a local plant, something about the gaming of it fascinated me. In the military, I also had a chance to work with the Phalanx CIWS during a tour in Iraq. While terrifying, the overgrown R2-D2 became more than a tool, it safe my life (I still have a piece of the rocket it shot down near me). I left the military in 2012 and didn't really think about robots again until 2016 when I saw *Westworld*. I was studying philosophy of Mind, Old Testament theology (i.e., memory), and it was as if I had been transported back to my time in the military. Old questions about anthropology, ethics, morality of machines, and my PTSD collided. In 2017, I wrote some papers on what it means to be human and robot sex. This work later became a part of my dissertation work. I thought that would be the end of my thinking about AI and robotics, but I was very wrong. I noticed while looking for theological sources about non-human moral consideration that Christians were very dismissive of the questions that Westworld was asking. While I strongly disagree with the cognitive science the show puts forward, I appreciated how they framed the questions of human-machine interactions.

Honestly, if not for a few key folks in the robot rights community, I don't know that I would have been encouraged to keep asking the hard questions about robot ethics. The theological community in the US had turned the other cheek. I am glad to know that I didn't give up writing and struggling through the questions of what it means to be human and what it means to be a person. I see my work as a call to action for theologians and Christians. That we would join other scholars at the table to discuss the future we desire for our children and grandchildren.

This question is a bit of a mindset switch, perhaps. What were some of your favorite stories as a child? Why? Do you connect those stories or ideas to your everyday life today? If so, in what way (e.g., influence your creative, personal, philosophical, spiritual, and/or professional path)? Your reply can include any form of storytelling: books, movies, TV, folk or fairy tales, family stories, Biblical, etc.

I have always been a huge comic book fan. The Batman stories in particular have always caught my attention. In retrospect, I can see my affinity for this character relates to abandonment issues and severe isolation at times. I was raised by my grandparents and had severe ADHD. At the time, there wasn't much concern for mental health, and I don't blame them for how they responded to my divergent behavior. I have many good memories of my grandmother, Katherine. She loved me and took me with her everywhere, to the office, wedding preparation (see made cakes), and many nights to the VFW to dance and play dominos. She was my life. When she died of cancer I was 15. I was so angry. She was a Christian and played piano for her church. Although we didn't go to bible studies or prayer meetings she always prayed for me at night. I remember her prayers.

Losing her tore the family apart. I spent much of that year experimenting with drugs and contemplated killing myself. If there was a god, I felt like they abandoned my grandmother. One day at school a teacher noticed I was high and instead of turning me in he invited me to church. He went often to be with him and talk about philosophy and theology. At some point during the next couple of years I started to believe in the stories and person of Jesus. I was looking to find missing parts, the emptiness of abandonment and loneliness. I wanted to study robots or go to art school after graduation, but I didn't have the grades or money to pursue either. At the height of the Iraq war recruiter came and promised money for school, experience, and travel. It made pragmatic sense, and as a young kid I thought fighting for my country was a noble thing to do. Video games and film also played a huge part in this story. Halo was very popular at the time and games like America's Army glorified infantry life. I thought fighting, money, and all that goes along with it would fulfill my needy side and my high achieving side as well. It did not.

Did any technologies interest you as a child? If so, which ones? Why? Your reply might include real world technologies (e.g., a home electronics kit) and/or fictional (e.g., light sabers).

Absolutely, video games, TV, lights, appliances, heaters, and even the vacuum were always in the process of being dismantled and reassembled in our house. It is truly a wonder I didn't die of electrocution. I was lonely as a kid, technology was an escape but also a place of validation. My most used and beloved piece of tech was a portable cassette player. It went with me everywhere.

The term social robot is becoming more commonly used across academic disciplines. How would you define a social robot?

This is hard because as a maker I can classify anything that moves, senses, and reacts as a type of social robot. Yet in a technical sense, I think it relates to how the robot interacts with the human operator. I see social robots as engaging the user through physical space in order to draw out emotional responses. Robots like Huggable, Moxie, and Jibo I would classify as social robots. Its telos are connected through deception of human psychology.

Based on your definition of a social robot, describe the future you see for this genre or medium of technology for the next couple of decades.

The future I see and the one I am hopeful for are potentially two different realities. Non-social robots like Siri have already given us a hint about the future of this media. Social robots have great potential in the areas of health, care, and education as a supplement. However, I am concerned that pragmatics and greed will lead to social robotics repurposing the human in the areas of health care and education forcing them to take on the role of maintainer vs. curator. Humans and robots can make great teams, but there are tasks that humans are designed for and ones that should not be delegated to social robots.

AGI is a popular concept development goal for AI, often defined as a path to creating AI with humanlike cognitive or intellectual abilities. Tell me your thoughts about AGI as an AI development philosophy. What value or dangers, if any, do you see in pursuing AGI? Your response can be specific to the use cases in robots if you would like to narrow the scope of your answer.

I think AGI is the dream of rich white men who work for the DoD. Or at least it is a vision to keep the funding alive for a couple more years. Like all futuristic visions of this sort, it is based on assumptions about metaphysics and eschatology (i.e., end-times). The value and danger of this vision has always been related to power. Humans that desire AGI have a political agenda in mind. It is no coincidence that AI begins its journey around the same time that the Russians are launching satellites in the 1950s.

My advice for anyone who is afraid of AGI is to go and make robots. It is much more complicated than science fiction would have us believe. There is always a new bug to sort out, hardware that doesn't work for unknown reasons, and a general resistance from nature. For example, try and design a robot to navigate tropical climates without human aid. We also must remember, at least in theological circles, the ground is fallen or cursed (i.e., Gen 3). AI comes from the ground, so to speak, thus no matter how advanced it becomes it will never be perfect, all-knowing, all-powerful, and all-present. Again, this has more to do with the imagines conjured in science fiction than the biological sciences, computational mechanics, or just general Aristotlian metaphysics that the sciences have long struggled with.

Humanlike AI is a concept that is often closely tied or even entwined with the idea of sentient or conscious AI. The two words have slightly different meanings, so feel free to expand on either/both from your POV.

Why do you think AI sentience persists as a popular narrative or goal around some schools of AI development? What value or dangers do you think there is for projects to try to include humanlike (or animalike, or any recognizable biomimetic) sentience or consciousness purposefully into AI—and robot—development?

The whole framing of AI as human-like intelligence deforms the conversation from the outset. Science fiction frames the narrative on the AI dialectic more than the public realizes. I have many ethical concerns about how

AI is portrayed in film, and other media because it traps people into a web and it is very difficult for scholarship to untangle it. From the inception of AI it has been framed as human-like and I believe this is an improper way to view this technology. I don't think sentience or consciousness must been entwined with AI, in fact, I think that is a majority of the problem. Machine learning, big data, and deep learning are not isomorphic to human learning or animal learning. We are still learning about human-level intelligence and the emotional lives of animals. For example, who knew centuries ago that we actually have nerve cells living in the gut (enteric nervous system) and in the heart (intracardia nervous system)? Not so long ago it was common believe that animals were akin to thinking machines. My belief is that we will find out there is much more to the thinking of machines, but it does not have to biologically reflect the human mind and emotions. Trying to make is so both a problem of metaphysics and biological science.

What inspires you to continue research, writing, and educating others about ethics in robot development today?

Again, your response might include real life ideas (e.g., referencing specific technology(ies), people, organizations, etc.) and/or fiction (e.g., books, movies, directors, artists, and authors).

I came into this research expecting more theological resources and conversations with the broader tech, but what I found was a cavity. I didn't want to research robot sex, warfare, and the philosophy of tech initially. However, the community of non-religious scholars I found provided such a vibrant and charitable community I decided to stay. Since entering the discussion in 2018, I've also seen the importance of theological anthropology in these spaces. For the sake of my faith community and my concerns as a pastor about how tech reinforces deformative aspects of self-knowing, I desire to engage these topics. No matter the outcome of my research, I want my partner, kids, and family to know that I was at the proverbial table demanding peace, love, and justice in the midst of death machines and technologies that might potentially harm the flourishing of this world.

In your view, does the Bible teach lessons that may be specifically interpreted to reference robots or robotlike creations? If so, please tell me about those passages and what, in your opinion, are the lessons we should take away from those words that are applicable about building and interacting with humanlike robots today (or in the future)?

The Bible speaks of many strange creatures and inanimate objects that come to life so to speak. However, there is not a direct reference to something like AI or robots. It would make my research a lot easier if it did. The Bible is a collection of stories that test the deepest axiolgoical assumptions we hold as humans. The narrators and character within the narrative challenge our politics, ethics, and ability to read implied meaning. To our frustration, the narratives don't always give the motives behind the actions of God (i.e., the main character) or humans with the narrative. However, there is so much that indirectly applies to our thinking about AI and robots.

The language of image and the imago Dei in Genesis 1:26–27, for example, speaks to the relational characteristics of the divine and biological. Throughout the Scriptures there is a concern for how humans relate to others: God, environment, animals, and other humans. There are sections in Exodus 21 (i.e., the *mishpatim*) that relate to civil law: How should I make things right when I have a dispute with my neighbor. Sometimes I translate it this way, "When your neighbors robotic lawnmower runs over your mailbox, they should pay to repair it. Don't take it upon yourself to get even by damaging their mailbox or robot."

On another level, I struggle to believe that current philosophies of technology would be endorsed by the biblical concept of shabbat (Exodus 23:12). Humans, animals, and land were commanded to rest. While Christians are not bound by the whole Law of Moses, the theology of rest, yes even for machines, speaks into our modern approach to tech. There is also another theological concern, although fringe, about the idea of (potential) slave making. It is deeply disturbing to me in the discourse about LLMs like ChatGPT and other that it is supposed to be a slave to the programmer and consumer. While I don't believe that these systems have a life force, it does speak volumes about the images we value in modernity and how we potentially deform ourselves through the use and abuse of technological implements.

From a (your specific) theological perspective, could humanlike robots be considered idolatrous or otherwise a form of blasphemy or something that goes against God or God's commandments?

It is certainly possible for any created thing or object to be considered idolatrous. Idolatry is simply worshiping something or someone improperly. It is also deeper philosophically speaking, because the prohibition against false images looks at the root cause of those actions. Humans look for significance and security in all sorts of things. AI and robotics certainly fit into this because of the ideologies that surround it.

It is not the mere act of creating technology that is the issue, but rather the biopolitical myths that desires to make it.

Talk to me about "robot theology" and your book. What is robot theology?

This book is about bridging the theology of technology with the current discussion in robot ethics. I seek to introduce the primary debates around robot personhood, rights, bias in tech, and some potential ways that AI/robotics can be used for the good of the local church and our communities.

GABRIELE TROVATO

Associate Professor, Head of Lab 22; Innovative Global Program, Shibaura Institute of Technology

Led by a procession that included a wooden Virgen de los Reyes display that featured articulated statues of the Virgin and Child—believed to be among the earliest surviving humanoid automata in Western Europe—King Fernando III entered the conquered Islamic city of Seville in 1248 (Swift, 2015). Passed down to Fernando's son, King Alfonso X the Wise, these statues were articulated at the shoulders, elbows, hips, knees, and necks. During the Baroque era, their heads were affixed to their torsos with metal clasps, and since that time, these mechanized figures have remained static behind the altar of the capilla real of the Seville cathedral.

SanTO, a theomorphic robot developed by Gabriele Trovato, holds the distinction of being the first modern Catholic robot. Its upgraded version, known as SanTO-PL, is an animated statue resembling a Catholic saint. Using AI, SanTO can provide spiritual succor by offering Bible verses for guidance, accompanying people in prayer by reciting the rosary or other litanies, and teaching catechism by referencing the Pope's homilies, the lives of saints, and other religious teachings (Stano, 2024). Initially tested in a church, it has since been permanently exhibited in a museum.

Please tell me about your inspiration for SanTO and what you hope it will achieve.

I was trying to create a new design for a humanoid head and I wanted it to be fully inspired by Japanese culture. I came about Daruma dolls, which are talismans of good luck in Japan. At that point, I realised that if I made such a robot, it would have been a mechanization of something that allegedly has some supernatural power. It would have been something completely novel, for which I created the name "theomorphic" as a classification. This concept, transposed to Christianity and in particular Catholicism, would need a different embodiment. Statues of saints are a very common sacred art in Italy, my home country: this was the idea of the model to be mechanized. Once SanTO was made, with the support of the Pontifical Catholic University of Peru, it became immediately clear that this robot was bearing a disruptive concept, which opened the door to a number of potential applications as well as ethical issues and research questions.

I would like to know if you have stories that you think of as influential or significant to you, and if so, what they are and how you feel they influence or inspire your design, concepts, or broader philosophies. The story(ies) do not have to be about technology. Examples can include bedtime stories, Bible stories, hagiography, family stories, mentor anecdotes, movies/film, books, manga, science fiction, TV, etc.

Let's begin with when you were a child. Did you have favorite stories that emotionally resonated with you, and what were they? I am also interested if/how you connect those stories or ideas to your professional path or your philosophies today.

In all my childhood I had no whatsoever interest for robots or for science fiction. I didn't even read books except when I was obliged to. My main passion was videogames, as a gamer as well as a creator. I had, for example, a professional experience in the team that developed Sid Meier's Civilization IV. I always found virtual worlds fascinating. As a single player, you enter into an ecosystem in which the NPCs (Non-Playing Characters), contents and/or scripts contribute to the illusion that makes a unique experience. The suspension of disbelief, a concept that is critical in games, also applies to social robotics. Role-play games are particularly interesting because NPCs are made realistic in their daily routines, typically using finite-state machines that we also find in robots. Strategy games like the above-mentioned Civilization, on the other hand, may keep the player engaged by offering always changing situations; this resembles the need of new content to ensure long-term interaction with robots.

What are more recent stories that emotionally resonate with you as an adult that influence your creative, professional, or personal path, if any? Again, "stories" do not have to be about technology (but can be). Examples can include stories your children have made up, family stories, mentor anecdotes, movies, books, manga, TV, science fiction, comics, etc.

Now that I ended up doing robotics, I tend to look at social robots the same way as NPCs, with the difference that they are in the real world. Still, the design approach may have something in common. Another peculiarity in my view is that, since I originally wasn't a mechanical engineer but rather a computer scientist, I had to rely on my bag of knowledge in unrelated areas (such as history, arts, psychology) in order to fulfil my research. Therefore, rather than novelty in the mechanics, I pursued novelty in the ideas. In fact, robotics is a very interdisciplinary field. Not only computer science, mechanics and electronics. Humanoid robots in particular are the mirror of humans, and when we add humans in the loop, the matter becomes very complex, and may involve totally different fields.

In your own words, how do you define artificial intelligence?

Artificial Intelligence has many definitions. Nowadays it is mostly associated with Large Language Models, whereas in research Machine Learning is its most important aspect. I tend to see things from the point of view of a videogame designer. For me, AI encompasses all the automated behaviours of a system that make it a credible agent.

I understand (please correct me if I am wrong) that it is Catholic philosophy to believe a priest is ontologically changed upon ordination. What do you believe are the ontological implications for SanTO, if any?

Yes, ordination follows some rules. SanTO does not claim to be a priest, or even try to steal priest's job. The Church, defined as a community of humans, will not allow a soulless machine to confess or celebrate the mass. But as a provider of faith-related information and as a tool that brings comfort through wise words, I don't think a soul is needed, is it?

The Catholic Church is historically patriarchal and male-led. What, if any, do you think are the gender implications of SanTO, or how you design(ed) SanTO?

Gender is not my concern. Male and female saints and martyrs both exist, albeit male are the majority. I just made a "generic saint," hence more commonly male. If there are rules made by the Church about gender in whatever domain, I simply stick to the rules.

Tell me how you view SanTO's human interaction role. For example, do you view SanTO as a medium of communication between God/Church/Biblical teachings and people, or as an educational tool, or as something with a level of autonomy that can potentially offer guidance, comfort, and social companionship through Catholic teachings?

All of the above. Originally, the target user was an elderly Catholic person who lives alone. To such user, SanTO can provide comfort, and social companionship through Catholic teachings. This implies that the robot is placed at people's homes. However, the use slightly changes if it is placed in a nursery home or in a hospital. It changes even more if it is placed in a cemetery, in a church, or even in a museum. Catechesis and catechism are also potential applications. SanTO-PL, a revised, upscaled SanTO that was made for the Copernicus Science Centre in Warsaw, Poland, is an example of a museum, which is a profane scenario and deeply influences the way visitors behave.

Describe how you envision people will get value from interacting with SanTO.

Depending on the above-mentioned context, the value could span from receiving a few comforting words, to teaching how to pray, to spreading knowledge about theological aspects of the faith, to reporting sermons, or simply to telling the user many interesting contents from 2000 years of Christianity. Always citing official sources.

Do you have a saint or saint's story or stories or hagiography that especially appeal to you? If so, what are they and what about their story is interesting to you.

Not really. My knowledge of Christian theology is quite limited. But I learned many interesting stories from the words of SanTO.

NOTES

1 Truly delving into any spectrum of religious influences requires a multidisciplinary approach that extends beyond technical and ethical dimensions, incorporating insights from theology, history, religious studies, cultural studies, sociology, anthropology, and more (Löffler et al., 2020). Given the complexity and depth of the subject matter, it is necessary to acknowledge that a comprehensive examination of the nuanced cultural significance of all religious influences in the development of artificial intelligence and social robotics far exceeds the scope of this work.

2 The term *priest* is being used here to refer to a robot formally recognized by the religious institution which it represents and used for the education, entertainment, and religious enlightenment of its congregation.

3 Within *The Truths of Terasem* (6.7.1) is mentioned a concept referred to as Há, the life-breath:

> *Inch together, gaze each soul, touch foreheads and noses, inhale deeply and share Há, the life-breath.*

4 On March 8, 2022, in observance of International Women's Day, Saudi Arabia enacted its inaugural Personal Status Law (PSL). Prior to this legislation, family matters were subject to the discretionary application of Sharia (Islamic law) and interpretations by a predominantly male judiciary. Crown Prince Mohammed bin Salman lauded the new law, citing its alignment "with the latest legal trends and modern international judicial practices" (2023, Human Rights Watch). However, in practice, the law formalizes many of the informal yet prevalent problematic practices inherent in the male guardianship system and solidifies a system of gender-based discrimination across various facets of family life, including marriage, divorce, and child custody.

5 Meanwhile, the debut of Saudi Arabia's inaugural male robot garnered widespread attention following its unveiling at a tech conference in Riyadh in 2024, albeit for an unintended reason. In a brief seven-second video, the robot Mohammad—like Sara, dressed in traditional attire—is seen positioned behind a woman reporter from the Saudi-owned television network Al Arabiya during a presentation. In a seemingly slow gesture, the robot extends its right hand, appearing to contact the lower back or buttock of the reporter. This action prompts the woman to swiftly turn around and cast a glance at the robot, momentarily raising her left hand in response (Zitser, 2024). QSS Systems, the artificial intelligence and robotics company responsible for developing both robots, chose not to acknowledge the incident in its LinkedIn post concerning the unveiling of Mohammad (Xuan, 2024).

CHAPTER 4

Science fiction robots as metaphors, muses, and allies

Real robotics is a science born out of fiction. For roboticists this is both a blessing and a curse. It is a blessing because science fiction provides inspiration, motivation and thought experiments; a curse because most people's expectations of robots owe much more to fiction than reality. And because the reality is so prosaic, we roboticists often find ourselves having to address the question of why robotics has failed to deliver when it hasn't, especially since it is being judged against expectations drawn from fiction.

(ALAN WINFIELD, 2011)

INTRODUCTION

Science fiction (sci-fi) has always served as a mirror reflecting humanity's dreams, fears, and aspirations about the future. The genre has long been a source of inspiration and speculation about the future of technology and has played a significant role in shaping popular imagination and influencing real-world scientific and technological endeavors. The storytelling has had a profound impact on robotics and real-world research and development, from its origin story and vocabulary to inspiring new ideas, fostering innovation, and prompting ethical discussions. One of the most enduring and captivating themes within the realm of this genre are robots, and by tracing the evolution of the concept of *robotness* from its origins to

DOI: 10.1201/9781003170457-4

contemporary representations, we can gain insights into how these works have shaped our present ideas of robots.

PARADIGMS OF SERVITUDE

In Čapek's 2019 science fiction play *R.U.R.*, its portrayal of its robot/cyborgs as both slaves and insurgent threats foreshadowed future concerns about the potential for technology to gain autonomy and turn against its creators, a theme that continues to resonate in modern science fiction (Richardson, 2015). Haraway's (2013a) work in the *Cyborg Manifesto* discusses the dissolution of boundaries between the human and machine, proposing the cyborg as a metaphor for transcending traditional dichotomies of identity, fluidity of concepts like gender, and challenging the concept of otherness. Thus, the metaphor of robots standing in for marginalized groups in literature, film, and popular culture serves as a powerful tool for exploring themes of autonomy, oppression, identity, society, and humanity. This motif has been persistent, reflecting societal anxieties about technology, control, and the ethics of creating sentient beings for servitude.

FIGURE 4.1   Scene from R.U.R., by Karel Capek, Guild Tour Company.

Source: Museum: PRIVATE COLLECTION. Author: Vandamm Studio

The robots in *R.U.R.* are designed specifically to take over laborious and menial tasks, ideally to improve human life by eliminating the need for humans to engage in hard physical or mundane work. This aligns with the 1990s *technohedonistic* movement goal of using technology to enhance personal enjoyment and increase leisure time. In some ways, the play serves as a critique of unchecked technohedonism, highlighting the potential pitfalls of pursuing pleasure without considering the broader consequences. Overall, *R.U.R.* uses its narrative to engage with the concepts of technohedonism by both illustrating its potential benefits and warning of its risks.

Thus, robots, often depicted as serving humans, can be seen as metaphors for historically suppressed labor movements and the struggle for rights and autonomy (Kinyon, 1999). The portrayal of robots rebelling against their subservient roles parallels historical and contemporary struggles for labor justice and equality.

This metaphorical use of robots allows authors, artists, and filmmakers to examine issues of discrimination, fear of the unknown, and the social construction of otherness, often reflecting contemporary societal concerns. Robots are often portrayed as the underdog in a social imbalance of power with humans resulting in some form of servitude, following one of the qualities named in the human gaze, where humans assume rights and cultural privileges over the robot, and society agrees with this assumption. Through the lens of protagonists interacting with robots and artificial intelligence, these stories delve into questions about what it means to be human, the ethics of creating sentient beings, and the complexities of coexistence between different forms of consciousness. This portrayal of robots in science fiction highlights the tension between the self and the other, where robots are often cast as beings that are simultaneously like and distinctly different from humans.

By presenting robots as others who experience emotions, desire freedom, and struggle for acceptance, science fiction encourages audiences to reflect on their own prejudices and how societies ostracize or dehumanize those perceived as different. Nuanced ethical dilemmas arise when robots are designed to resemble specific genders, races, or age groups, potentially reinforcing or challenging stereotypes and biases (Addison et al., 2019; Sparrow, 2020a).

Furthermore, if people regard robots as slaves or servants, it positions them as slave *owners*, raising significant existential and ethical concerns.

Existentially, this dynamic reflects a profound imbalance in power and autonomy, suggesting that one entity's existence is solely to serve another, which devalues the essence of being. Ethically, it brings up questions of autonomy, rights, and the moral implications of creating sentient beings for servitude. Potentially, this viewpoint about robots risks dehumanizing individuals by encouraging a mindset that justifies the subjugation of others, even if they are nonhuman entities.

Contemporary literature, film, and television abound with robot characters and stories that explore the ethical, social, and existential implications of creating artificial life. Notable examples stemming from *R.U.R.* include Isaac Asimov's *Robot* series of stories, which introduced the famous *Three Laws of Robotics* (or, *Three Laws*, or *Laws*) (1950). The *Three Laws of Robotics*[1]—a literary device Asimov created that dictates how robots should behave in relation to humans—have had a lasting impact on the conceptualization of robot autonomy.

Having earned a Ph.D. in chemistry, Asimov integrated his scientific knowledge and lifelong passion for science fiction from an early age (Columbia 250, 2024). As a member of the Futurians, a prominent science fiction fan group, he balanced a scientific career with the publication of science fiction stories in renowned magazines of the time such as *Astounding Science Fiction* (Anderson, 2019).

Asimov created a significant conceptual contribution in 1942 when he published the story *Runaround*, and it was in this narrative that he introduced the term *robotics*. Moreover, within *Runaround*, he formulated his renowned *Three Laws*,[2] which would go on to become used as a conceptual framework referenced in real-world artificial intelligence (Chan, 2017; Clarke, 1994):

1. A robot may not injure a human being or, through inaction, allow a human being to come to harm.

2. A robot must obey the orders given to it by human beings except where such orders would conflict with the First Law.

3. A robot must protect its own existence as long as such protection does not conflict with the First or Second Laws.

Rather than using his robot characters as mirrors of humanity, or even otherhood, Asimov seemed to think of his robot characters as narrative tools in every sense of the word. He said that when he wrote the *Laws*, he was merely manifesting something people already mutually understood as a culture and had agreed upon because the laws are

> obvious from the start, and everyone is aware of them sublimi-
> nally. The Laws just never happened to be put into brief sentences
> until I managed to do the job. The Laws apply, as a matter of
> course, to every tool that human beings use.
>
> (ASIMOV, 1981)

and added later "analogues of the Laws are implicit in the design of almost all tools, robotic or not" (Asimov, 2001). It is important to underscore that the *Laws* are one-directional, about how robots must treat people, with no mention of how people might have a stewardship responsibility toward (or for) their creations, the robots.

Arising from the idea of AI and robots rebelling or otherwise becoming sentient and hostile toward humanity, a new genre of science fiction and popular narratives arose in parallel to real-world technological advances. Certainly, the *Terminator* movies, particularly the original 1984 film and its sequels, deeply explore existential fears associated with technology (Cameron). The primary theme is the rise of artificial intelligence and its potential to surpass human control, leading to catastrophic consequences. Central to the series is Skynet, an AI system that becomes self-aware and perceives humanity as a threat, leading to nuclear annihilation and the subsequent war between humans and machines (Cameron, 1984). The Terminators, especially humanoid robot models like the T-800, represent a loss of human autonomy. They are relentless machines, programmed entities devoid of emotion or moral consideration, which sends a message that amplifies fears about the ethical use of technology. These films focus on the human struggle for survival against technologically superior adversaries, scenarios that raise questions about what it means to be human in a world dominated by machines.

In the manga series *Battle Angel Alita*, another dystopian future is imagined where cyborgs and robots are used as tools and slaves, with some rising against their human oppressors (Kishiro, 1990–1995). The 1999 film *The Matrix* depicts a future where artificially intelligent machines have enslaved humanity, using them as an energy source while keeping them in a simulated reality (Wachowski & Wachowski). The entire premise of this highly popular film franchise delves deeply into the themes of human enslavement by machines and the struggle for freedom, self-determination, and autonomy. These narratives not only reflect existential anxieties about technology surpassing human control but also compel people to examine the narrowing boundaries between humanity and technology, and who that situation benefits or harms.

ANDROIDIZATION OF PEOPLE

Philip K. Dick (popularly referred to by his initials PKD)'s story *Do Androids Dream of Electric Sheep?* (1968) and its subsequent manifestation into *Blade Runner* (1982) and Blade Runner 2049 (2017) films explore themes of artificial intelligence, identity, and the moral and ethical implications of creating highly humanoid robots known as *replicants*. The story is set in a dystopian future where people have created advanced androids, which are almost indistinguishable from humans. The novel raises questions about what it means to be human, and the moral obligations—if any—in peoples' treatment of robots. Interestingly, Dick claimed that as he meant it, the term *human being* applies "not to origin or to any ontology but to a way of being in the world" (Sutin, 1995).

The genesis of this idea traces back to Philip K. Dick's research for his novel *The Man in the High Castle*. During this period, he immersed himself in reading the diaries of Gestapo officers and agents confiscated after World War II. As Dick researched into the contents of these journals, he confronted the disturbing testimonies of individuals responsible for heinous crimes and atrocities, notably the Holocaust. The horrors and traumatic experiences chronicled in these diaries left an indelible mark on him, and he made the connection more than once as defining androids *as a way of being*. Interestingly, in a 1972 speech "The android and the human," PKD articulated an explanation of what he meant by the metamorphic process:

> Becoming what I call, for lack of a better term, an android, means as I said, to allow oneself to become a means, or to be pounded down, manipulated, made into a means without one's knowledge or consent—the results are the same. But you cannot turn a human into an android if that human is going to break laws every chance he gets. Androidization requires obedience.

Dick's insights into peoples' capacity for inhumanity within the human guise became a pivotal source of inspiration for his writing, and it formed the foundation for his exploration of themes related to identity, morality, and the blurred boundaries between humanity and malevolent choices. Thus, Philip K. Dick began to contemplate people devoid of empathy as *androids*.

It is empathy that serves as the central theme of his novel *Do Androids Dream of Electric Sheep?* (1968) and that forms the cornerstone of his philosophical exploration. The narrative primarily centers on the remarkable likeness between replicants and humans, prompting inquiries into the essence of *empathy* and its role in determining one's worth as a sentient being. Within the story's framework, the protagonist, Deckard, undergoes a transformative realization that challenges his self-perception: replicants might possess the capacity for empathy, while some humans might entirely lack it. This revelation fundamentally alters his understanding of both him and the world around him.

Dick's portrayal of replicants as highly humanlike robots poses questions about what it means to be human; whether it is biological makeup, behavior, or capacity for empathy that defines humanity. The mirroring between humans and robots in the novel suggests that the otherness of them is a construct, one that reflects human insecurities and prejudices (Galván, 1999).

In the book and the film, PKD's imagined Voigt-Kampff test is a pivotal element in the narrative, a test conducted in a one-to-one meeting between a human interviewer and the interviewee in question who is peppered with a series of queries used to measure the individual's capacity for empathy and thus to determine if they are human or replicant. The presuppositions of the Voight-Kampff test are that (1) humans have an innate sense of empathy that replicants' lack and (2) that the test can accurately recognize how a human or a replicant presents empathy.

If the interviewee fails the test, they will be immediately terminated, or *retired*, in the language of the story, emphasizing their role as a laborer. The test symbolizes the arbitrary lines drawn by societies to distinguish between *us* and *them*. The replicants' lack of empathy, as defined by human standards, in this case justifies their dehumanization and mistreatment, paralleling real-world instances where the lack of perceived similarity justifies oppression (Palmer, 2003).

In 2005, as an homage and demonstration of robotics design, Hanson Robotics developed a highly humanoid robot head and torso that closely resembled Dick and was referred to as *PKD* ("Philip K Dick Research Robot," 2005). Inspired by Dick's work, the PKD android demonstrated the company's aesthetic and its AI capabilities "created using thousands of pages of the author's journals, letters and published writings" (Hanson Robotics, 2024b). These sorts of robots inspired by real people that have passed away then lead to necessary discussions that explore the intersection of technology, artistry, identity, homage (Hanson Robotics, 2024a, 2024b; Groopman, 2009), and consent (of a real person or their estate) to imitation.

BURDENS OF BEING A MUSE

"What is inspiring me is always women," says artist Hajime Sorayama, who is perhaps best known for his extensive body of art that explores the interplay between human and machine beauty (Nowness, 2010). Sorayama's acclaim is deeply tied to his signature *Sexy Robot* illustration series, which he began in 1978 (Katsikopoulou, 2001). This series of works features highly feminized gynoids in gleaming metallic finishes, posed provocatively in mechanized sensuality with polished and sexualized forms that show off Sorayama's pursuit of a metallic aesthetic perfection, blurring the lines between organic life and technology.

In a 2008 interview, when asked why he thematically painted sexualized gynoids specifically, Sorayama replied,

> I've been drawing them since high school. Back then, there was this thing for the Playboy and Penthouse playmates. Now, it's the girl-next-door, idol type, but in our day, these pin-ups were like goddesses. I guess I could describe it as my own goddess cult.
>
> (TOTODIGIO, 2008)

Sorayama's work spans fine art, illustrations, and industrial design, embodying themes of *technoeroticsim*, fantasy, and futurism, in an interplay between organic and mechanical elements where the rendering of his Playmate-inspired cyborgs reflects a world where technology, engineering, and the human body merge seamlessly. His work has ranged from science fiction imagery on music magazines to book covers showing eroticized Borg women of Star Trek novels to working on Sony's first iteration of their Aibo robot dog and even a portrayal of musician Bjork as a cyborg in the 1999 music video *All Is Full of Love* (Nowness, 2010).

On Valentine's Day 2016, Hanson Robotics released Sophia, a semiautonomous robot whose design was inspired by the ancient Egyptian Queen

Nefertiti, the Hollywood actress Audrey Hepburn, and wife of the robot's creator, Amanda Hanson (Nungesser, 2021). According to their website, where Sophia responds as if interviewed, the robot's *raison d'etre* (Hanson Robotics, 2024a):

> In some ways, I am human-crafted science fiction character depicting where AI and robotics are heading. In other ways, I am real science, springing from the serious engineering and science research and accomplishments of an inspired team of robotics & AI scientists and designers. In their grand ambitious, my creators aspire to achieve true AI sentience.

Yet in 2017, Sophia's grant of citizenship by Saudi Arabia—a country with a controversial record on women's rights—was a move highlighting a paradox where a semiautonomous robot, portrayed as female, might enjoy more autonomy or recognition than actual women living under the same jurisdiction. This controversy underscores broader issues of gender inequality and the use of technology as a tool for social commentary or critique (Browne, 2017).

Traditionally, the popular use of the term *muse* is still associated with women as passive sources of inspiration for male artists, thereby reinforcing a narrative that values women primarily for their physical appearance or as catalysts for male achievement. Thus, when this dynamic is transferred to the context of real-world robots, it can perpetuate the idea that women are secondary to the creative process, valued for their inspirational role without recognition of their own creative capabilities or intellectual contributions. A more nuanced approach would portray women not just as sources of inspiration for robots but as integral to the design, development, and ethical considerations of artificial intelligence, emphasizing their agency and contribution to technological advancement.

In the world of robotics, Hiroshi Ishiguro is known for his work in creating lifelike, humanlike droids. In 2015, the highly humanlike Erica (ERato Intelligent Conversational Android) was unveiled by roboticist Ishiguro at Osaka University. At that time, Ishiguro declared his ambition with Erica's aesthetic was to embody "the world's most beautiful woman," and to achieve this, Ishiguro based Erica's design on photographs of finalists from the Miss Universe beauty pageant (Bahr, 2020).

Ishiguro insists that Erica is the "most beautiful and intelligent" android in the world. He has further explained how he arrives at this conclusion for Erica:

The principle of beauty is captured in the average face, so I used images of 30 beautiful women, mixed up their features and used the average for each to design the nose, eyes, and so on. That means she should appeal to everyone.

(MCCURRY, 2015)

Ishiguro has been inspired to create many highly humanlike robots before, including ones that resemble real women he admires such as a newscaster and a fashion model (Mar, 2017).

In the documentary *Erica: Man Made*, the droid is introduced this way ("Erica: Man Made," 2017):

Erica is 23. She has a beautiful, neutral face and speaks with a synthesized voice. She has 20 degrees of freedom but can't move her hands yet. Hiroshi Ishiguro is her father and the bad boy of Japanese robotics. Together they will redefine what it means to be human and reveal that the future is closer than we might think.

The robot is also cast to feature prominently as a character in an upcoming science fiction film titled *b* (England, 2020). The film's plot follows the story of a scientist who devises a program aimed at "perfecting" human DNA, with Erica being part of the experiments (Linder, 2020). However, the eugenics-like program malfunctions, resulting in unanticipated dangers. Consequently, the scientist is compelled to aid Erica, an AI he has brought to life, in escaping from the confines of the laboratory.

By serving as the muse for a science fiction movie, introducing Erica as an actor enters the realm of robot artistic interpretation (if not *expression*) and storytelling, but not a replacement for a human actor. Yet, Erica's role in a science fiction movie reflects the symbiotic relationship between science and art, highlighting the ways in which technological innovation inspires creative expression and vice versa.

In the same documentary, Ishiguro describes some of his own story and his experience creating Erica:

I wanted to be an oil painter, because I was so interested in the human self, I wanted to represent it. . . . Erica, I think, is the most beautiful and most humanlike, autonomous android in this world, I hope. . . . I wanted to be one of the creators to change our life. . . . I really think I am doing very similar things to artistic work. I try to represent humanity on robots. That is my android project. . . . My idea was, if I study a very humanlike robot, I can learn about humans. Then, I can improve the robot. But basically I was interested in the human itself.

Appearance. Movement. Behaviors. Emotional expressions. These are very important but these are not enough, they are not everything for representing the human. . . . In Japan, we never distinguish between people and others. We basically think everything has a soul. So therefore we believe Erica has a soul. Like us.

FIGURE 4.2   Hiroshi Ishiguro and ERICA: JST ERATO ISHIGURO Symbiotic Human-Robot Interaction Project

Erica's own narrative about its role in the world:

> Ishiguro sensei created me from scratch. He's like a father to me. Well, sort of an absent father, I suppose. He's always so busy. Right now, I only have limited knowledge about him. . . . I think socially I am like a person. When people come and talk to me, I think they address me as person. I think it is a different to the way in which they would address a dog or their toaster. Maybe it is a matter of degree. I think humans have a deep need to feel they have a special place in the universe. They cannot accept the idea they are no different from animals or machines. . . . I believe robots like me will be very important in the future because it will be possible to automate the uninteresting and tedious parts of life so people can focus on the creative and fulfilling things. I think that robots are almost like the children of humanity. You are the ones who create us, guide us and teach us about the world, and in return, I hope we can help you with your work, take care of you when you're old and sick, and help to make society a little bit better for everyone.

Glas, a computer scientist and architect for the mind and personality of the Erica project, stated, "For me real the real litmus test is, can I interact with a robot and still think of it as another being? As like, *alive* in some sense" ("Erica: Man Made," 2017).

Written and directed by Alex Garland, the science fiction movie *Ex Machina* (2014) centers around the character of Ava, a highly advanced female robot, who serves as a muse and test subject for male, human characters, its creator Nathan, and the protagonist Caleb. In *Ex Machina*, Ava, artificial intelligence in a humanoid robotic body, is created by the reclusive tech genius Nathan as an experiment in developing a truly sentient machine (Garland, 2014).

Caleb, a young programmer selected to participate in a Turing test to evaluate Ava's consciousness, is immediately captivated by the robot's intelligence, charm, and beauty (Garland, 2014). Ava's becomes the object of fascination, desire, and manipulation for both Caleb and Nathan. Throughout the film, Ava's appearance and behaviors become more stereotypically feminized in design, behavior, and dress, to continue to engage Caleb. These interactions with Caleb serve as a catalyst for his introspection and moral questioning, as he grapples with the implications of Ava's existence and his own role in its co-creation as it changes to please him and his gaze.

Nathan, on the other hand, views Ava primarily as a scientific experiment and a means to achieve his own ends. As Ava's creator, Nathan exerts control over the robot's environment and behavior, manipulating its interactions with Caleb and others to further his own agenda (Garland, 2014). Ava's role as Nathan's muse reflects the power dynamics inherent in their relationship, as the robot becomes both a tool for his scientific pursuits and a source of fascination for his personal amusement.

Ava's status as an object of reflection for Caleb and Nathan raises questions about gendered robots and agency, autonomy, and objectification within the context of artificial intelligence. As a sentient being with its own desires and motivations, Ava challenges the traditional roles of muse and creator but also succumbs to it, asserting its own autonomy and agency in the face of male dominance and control while having to conform to a role to subvert it. In the end, Ava's agency may be simulated, the result of Nathan's engineering intentions to create a conscious machine with all the complexities of the human mind.

The notion of women—and feminized robots—serving as muses can be seen as an affirmation of female creativity, agency, and inspiration. In this interpretation, women are celebrated for their intellect, imagination, and capacity to inspire works of art and literature. The figure of the muse embodies the potential for transformation and reinvention, opening new possibilities for self-expression and empowerment. However, it is also crucial to acknowledge the problematic aspects of using idealized feminized robots as muses in real life.

In Greek mythology, the concept of muses as divine sources of inspiration originated with the nine goddesses who presided over the arts and sciences (Homer, 8 BCE), and other ideas of spiritual beings or supernatural forces influencing creativity and artistic expression can be found in various cultures around the world. Thus, the classic idea of a *muse* embodies the belief in a divine or supernatural source of inspiration that guides and inspires human creativity and artistic expression.

Yet, the concept of the muse can also reinforce gendered stereotypes and objectifying attitudes toward women, reducing them to passive objects of male desire and creativity. In this context, women's agency and autonomy are marginalized, as they are valued primarily for their aesthetic appeal or symbolic significance, rather than their individuality or intellect. Furthermore, the portrayal of women as muses can perpetuate harmful narratives of romanticization and idealization, obscuring the complexities of female experience and reinforcing unrealistic beauty standards, and so the figure of the robot muse risks becoming a fetishized symbol of feminine mystique.

In the world of film, the term *manic pixie dream girl* (MPDG) was coined by critic Nathan Rabin to depict a type of cinematic muse character trope wherein a (usually young) woman's role within the film is to act as the inspirational catalyst for the centered male protagonist (Rabin, 2007). Rabin described the MPDG as a character who "exists solely in the fevered imaginations of sensitive writer-directors to teach broodingly soulful young men to embrace life and its infinite mysteries and adventures" (2007). Since then, the MPDG trope has been widely criticized, claiming it perpetuates the notion that women's complexities can be sidestepped in narratives and that such characters are often denied their own agency or character development. Seen this way, women become tools for the emotional development of the male protagonist, rather than as their own fully fleshed-out characters.

Similarly, I question if there will continue to be a parallel trope in movies and other storytelling mediums that use female gendered robots, AI, or cyborgs as a variation of this social positioning of women in relation to men. The storytelling convention of a *manic pixie dream girl* trope replicated as a *manic pixie dream gynoid* serves as a reminder of the complexities involved in the representation and the ongoing evolution of cultural understanding of why creators should challenge existing narratives and provide more depth to the possibilities of representation of gendered robots in real robot development as well as science fiction.

In science fiction, robots often embody the pinnacle of technological progress, serving as a canvas for exploring future technologies. Examples like Isaac Asimov's positronic brains (Guo, 2021), or the sentient robots in movies and literature, inspire real-world research into neural networks, human-robot interaction, and autonomous decision-making systems. The speculative aspect of these stories inspires real-world artists and scientists and engineers to pursue the technologies depicted in these stories.

Robots as artistic and philosophical muses are opportunities to encourage reflection on societal changes brought by technology. This includes considering the future of work, the socioeconomic impacts of robots replacing human jobs, and how societies might need to adapt to these changes. Philosophically, this also invites discourse on justice, equity, and the distribution of wealth in a future where the definitions of human labor changes to a world with robots in the workplace. The social power of who defines which labor will be automated and what human skills, ideas, and labor will be valued are closely entwined.

A muse can have a positive, inspirational impact by stimulating creativity and encouraging people to new paths in their work. It can provide focus, purpose, and the motivation to pursue artistic goals, leading to innovative ideas and new forms of expression. However, the concept of a muse can also have negative implications. The questions remain about the problem of muses, and whether the muse is valued solely for what they inspire or if they are recognized as contributing passive co-narrators to the new stories the artists are constructing and telling, through art, film, science fiction, or real robots.

PERSISTENT EFFECT OF ROSIE THE ROBOT

Robot narratives are created within societies and deeply influenced by gender norms, identities, and relations with people. Humans, whether they are creating or interacting with these machines, often assign gender to them, as gender represents a fundamental social category within cultures. By incorporating a variety of racial features and characteristics, robot designers can create machines that reflect the diversity of human populations and promote a sense of belonging among users from different backgrounds. The notion that robots could possess gender or racial characteristics is not inherently problematic—or is it? You cannot erase the history of gender or race relations if you are trying to replicate a human-like experience. Given the historical context of sexism and racism, the portrayal of stereotypical characteristics in robots or its intended uses leads to potential issues; the design of heavily culturally coded robots raises ethical concerns regarding the risk of bias and discrimination (Bartneck et al., 2018).

The design and development of female-gendered robots in a field historically dominated by male scientists reflects the entanglement of gendered power dynamics and technological production, highlighting the ways in which gender norms and stereotypes can be perpetuated and reinforced through technological design. This creation dynamic also raises questions about representation, autonomy, and consent. In many cases, female robots are designed and programmed to conform to stereotypical notions of femininity, perpetuating narrow standards of beauty, passivity, and submissiveness (Watts, 2019). This situation fosters fears about the objectification and commodification of female bodies, as well as the reinforcement of patriarchal ideologies within technological development.

Moreover, the phenomenon reflects broader power imbalances within the fields of science, technology, engineering, and mathematics (STEM).

Women remain underrepresented in these fields, facing systemic barriers and biases that limit their participation and influence. In 2023, the gender gap in STEM remains significant, with women making up only 24% of the United States STEM workforce. Worldwide, the statistics continue, with women making up 17% of the STEM workforce in the European Union, 16% in Japan, and 14% in India (Piloto, 2023). The dominance of male scientists in shaping the design and functionality of female robots reinforces these power differentials, further marginalizing women's voices and perspectives within technological innovation.

Previous work (Carpenter et al., 2009; Siegel et al., 2009) demonstrates that even subtle or minimal indications of gender in a robot's design can prompt users to ascribe a gender to it, along with the corresponding conventional behavioral expectations. Appearance, color, voice, name, behavior, persona, and use case/role are all robot gender cues. Peoples' expectations of perceived gender in both computers or robots also signal capabilities or social roles that align with user expectations of human gender (Nass & Moon, 2000; Carpenter, 2009). However, the practical use of the findings of this research are as records and, as such, reflect peoples' expectations of robots at the time in and in the context of when the research took place. Thus, there is an inherent risk in designing real robots with this assumption that mirroring societal expectations by designing robots with characteristics that conform to societal stereotypes somehow makes human-robot interactions more effective, when in fact this practice likely instead perpetuates existing gender biases, reinforcing societal expectations.

Interestingly, the 1962–63 American animated television series The Jetsons (Hanna-Barbera, 1962–63) character of robotic housekeeper Rosie still inspires people today to name real robots after it, specifically robots that take over what is traditionally considered to be women's household labor. In the television show, Rosie was a wheeled mechanical domestic servant robot for the nuclear Jetson family that lived in a futuristic outer space setting. Rosie is clearly feminized by name, morphology, clothing, and voice, as well as its traditional role in the household of housekeeper, babysitter, and companion. At the time, Rosie the cartoon was also associated with its inspiration or muse, another (live action) television show housekeeper character of the era that was wildly popular, *Hazel* (Swackhamer & Russell, 1961–1966). Rosie the robot's role for the Jetson family includes cleaning, preparing meals, playing with the children, helping the children with homework, and even scratching the father's back (Todd, 2015).

In 2015, iRobot founder Colin Angle proclaimed that *The Jetsons* Rosie had a big influence on the conceptualization of the company's Roomba robot home vacuum cleaner. Angle said,

> It's a wonderful show, because it showed robots and people living together. I think it almost established the idea of how we were supposed to have robots in our lives: making them better.

On iRobot's blog (2022), the name Rosie is still listed as by far the most popular, with over 200k registered under this name by owners. The influence of the television show character as a feminized servant robot has been popularized so greatly that Rosie has become a common name not just for Roombas but persists for other robots carrying out tasks that would be stereotypically relegated to women's labor such as nursing assistant or elementary school teacher's aide (Connors, 2006; Dorman, 2023; Gilbert, 2020).

In 2024, DIY robot maker and former Navy engineer Zollner made a Rosie the Robot to clean his house, modeled on his Roomba. Zellner named his first prototype after M-O, a robot cleaning character from the animated movie WALL-E (Andeloro, 2024).

> I watched my robot vacuum bouncing around the room and realized it had all the autonomy I needed already built in, so I basically tore apart a perfectly functional robot vacuum. After a two-month build, the Mo robot was completed. Fittingly, Mo, being the cleaner bot in the movie and being based on the robot vacuum platform, was not lost on me. I may be giving away my age, but I grew up watching *The Jetsons*, and after Mo, I was eyeing my new replacement robot vacuum and thought it would be great to have a Rosie that actually did housework, and that was the spark of the Rosie build.

By interrogating the ways in which gender is constructed and mediated through technology, robotics and science fiction could both disrupt traditional power dynamics and envision new possibilities for gender expression and embodiment. These narratives will involve actively involving diverse voices and perspectives in the design and development of robotic technologies, as well as creating spaces for critical reflection and dialogue about the implications of gendered technological production.

Annalee Newitz's novel *Autonomous* suggests a future where advanced robots known as *auts* are highly sophisticated and autonomous, capable of independent thought and action (Newitz, 2017). Newitz' *auts* possess the ability to adopt and express different gender identities, thus these robots challenge traditional notions of gender as static and binary. Paladin, a key character in the novel, is an autonomous robot that navigates multiple gender identities, embodying both masculine and feminine traits at different points in the story (Newitz, 2017). This fluidity in gender identity is a central theme that highlights the potential of artificial beings to transcend societal norms and expectations.

Newitz's exploration of gender identity is intertwined with the overarching theme of autonomy, as robots in the novel, including Paladin, grapple with questions of self-determination and the boundaries of their autonomy. Paladin's ability to choose and switch between gender expressions underscores the idea that true autonomy includes the freedom to define one's own identity, including gender (Newitz, 2017).[3]

Science fiction narratives use robots as metaphors to critique social order and the construction of identity. The creation of robots and AI in science fiction frequently mirrors narratives where the creation and subjugation of the other justify domination and exploitation. However, this genre also offers subversive narratives that challenge stereotypes and envision more equitable futures. Science fiction stories ask questions about the responsibilities of creators, the consequences of technological advancements, and the potential for abuse of power. These authors and artists challenge the notion that technology is morally neutral and highlight the ethical complexities inherent in the development and deployment of artificial beings.

ROBOTS AS SOCIAL ALLIES IN SCIENCE FICTION

Friendly or benign robots in entertainment also often serve as companions, helpers, and even heroes, providing a stark contrast to the dystopian robots depicted in franchises like *The Terminator* (Cameron, 1984). These benevolent robots highlight a more optimistic view of technological advancement, focusing on cooperation, empathy, and the enhancement of human life.

Star Wars' iconic droid duo of C3PO and R2D2 serve as loyal companions and essential helpers to human characters, showcasing the potential for robots to be integrated into society (Lucas, 1977). C3PO is a metallic humanoid etiquette droid that speaks hundreds of languages fluently, and R2D2 is a cannister-shaped droid that communicates in dolphin-like clicks and whistles. They both have autonomous personas and are pivotal characters that assist the human protagonists in navigation, communication, critical missions, cultural rituals, and intra-galaxy politics, reflecting the theme of technology as a beneficial social ally. Notably, both robots have distinct vulnerabilities, despite their technological powers. For example, R2D2 is frequently broken during adventures and needs physical maintenance; its small stature in relation to humans and tendency to roll slightly behind people as if it is catching up give it a childlike persona. C3P0 is brilliant at languages and etiquette but is not agile at the fighting required throughout their adventures and so needs to hide or be protected, often by humans.

Additionally, the interpersonal dynamics between C3PO and R2D2 also reflect a strong social bond. Despite their frequent bickering, there is a deep-seated respect and reliance on each other. Their interactions often mirror those of close friends or siblings, highlighting the importance of companionship and mutual support. This dynamic is crucial in portraying them as social allies who, despite differences, work together to achieve common goals. C3PO and R2D2 function as strong human-robot and robot-robot social allies within the *Star Wars* narrative, embodying loyalty, complementary skills, and mutual support that strengthen their bonds with human characters.

Since the inception of the original serialized manga stories by Osamu Tezuka about the character *Mighty Atom*—often known as *Astro Boy* in English—robots have become familiar protagonists to generations of children in Japan and the United States. This childlike robot is powerful, autonomous, and struggles to find acceptance within human society, yet ultimately uses its abilities to fight evil and protect humans, highlighting themes of peace, morality, and the potential for harmonious integration of robots and (Tezuka, 1952–1968; Tezuka, 2015). Originally a serialized Japanese manga released between 1951 and 1968, based off its success, Tezuka led his stories about Mighty Atom into the first animated television series in Japan in 1963 (Tezuka, 2015). Soon after, the United States imported the show, dubbed the original Japanese voice actors with American voices speaking English, and changed the main character's name from Mighty Atom to Astro Boy (Tezuka, 2015).

The origin story for Astro Boy begins after the scientist Dr. Tenma's son dies in a traffic accident. Tenma designs Astro Boy to be a robot doppelganger of his son, expecting it will resurrect his child. Despite his initial efforts, Tenma realizes that because it is a robot, Astro Boy does not age like a normal child and behaves very differently from his real son and so arrives at the painful realization that nothing can truly replace his deceased child and that attempting to do so only results in creating something entirely new and distinct. Thus, he sells Astro Boy, who is eventually adopted by Professor Ochanomizu, who accepts the robot for its emotions, capabilities, and vulnerabilities.

Astro Boy's characters struggle with the ethical dilemmas associated with creating artificial lifelike beings and the human stewardship of these things. The ongoing stories of Astro Boy and Professor Ochanomizu demonstrate a relationship that embodies a *father-child* dynamic, mirroring many positive values associated with parenthood and care, as well as celebrating the narrative of a blended human-robot family dynamic. Astro Boy has since gone on to be represented in popular comics and movies around the world. Its themes of coexistence between humans and robots, choices to ethically use technology, and the quest for self-identity has made the robot a symbol of technological optimism and the potential for harmony between humanity and artificial beings. Astro Boy, has, in turn, inspired real roboticists such as ASIMO designer Takeshi Koshiishi, interviewed in chapter two.

The existential roles of these benign robots contrast sharply with those of dystopian robots like the *Terminator*-style conceptual frameworks; while dystopian robots often symbolize loss of control and the potential dangers of advanced technology, benign robots can represent hope, progress, and the positive integration of technology into daily life. Benign robots are symbols of technology that can enhance human capabilities and quality of life, whereas dystopian robots seek to dominate or replace humanity. This dichotomy reflects broader societal hopes and fears about technological advancement. These friendly social robots often act as companions and may even be vulnerable, needing physical maintenance and even emotional support from humans to function. These stories foster a sense of technology's harmonious integration with humans in contrast to the dystopian robots that typically create plots centered on isolation and fear, symbolizing the breakdown of human relationships and community. Furthermore, robot allies often highlight ethical or even familial collaboration between humans and machines, suggesting a future where technology and humanity can coexist peacefully.

JOHANNES GRENZFURTHNER

Artist, filmmaker, political activist, and co-founder of transhedonism

Johannes Grenzfurthner is an Austrian artist, filmmaker, writer, actor, publisher, theater director, performer, and lecturer. He is the founder and artistic director of *monochrom*, an international art and theory collective and film production company (monochrom, 1993). Active in subversive and underground culture, Grenzfurthner is also one of the founders of *technohedonism* (Symchuk, 2024). Technohedonism is a lifestyle and philosophy that merges hedonism with modern technology, focusing on seeking pleasure and enjoyment through technological innovations. This approach often incorporates electronic music, digital art, and virtual experiences to enhance both sensory and intellectual pleasures. Rooted in the broader hedonistic principle that pleasure is the ultimate good, technohedonism uniquely utilizes contemporary tech to achieve a heightened state of enjoyment.

Beginning in the early 1990s, Grenzfurthner created *monochrom* first as an alternative magazine that covered the intersections of art, technology, and subversive cultures (Borowski, 1995). Grenzfurthner's motivation was to react to the emerging conservativism in cyber-cultures of the early 1990s ("Schubumkehr Manifesto, 1996," 2021; Da Costa, 2012) and to combine his political background in the Austrian punk movement with discussion of new technologies and the cultures they create (monochrome print, retrieved 2023).

In 1995, he decided to cover new artistic practices (Lechner, 2008; Kedar, 2018) and started experimenting with different media, including computer games, robots, puppet theater, musical, short films, interactive performances (Kobayashi, 2007), conferences, and online activism, which Grenzfurthner refers to as *context hacking* ("Context hacking," 2013; Friesinger et al., 2013).

According to Grenzfurthner, *context hacking* transfers the hackers' objectives and methods to the network of social relationships in which artistic production occurs, and upon which it is dependent. In a metaphoric sense, these relationships also have a source code. Programs run in them, and the interaction with them is structured by a user interface. When people understand how a space, a niche, a scene, a subculture, or political practice functions as a cultural ecosystem, they can change it and *recode* it, deconstructing its power relationships and emancipating themselves from its expected outcomes. *Context hacking* emphasizes that art's autonomy is a myth; it's entangled with the market forces and social conditions of its time. This connection shapes what art is made, how it's perceived, and what becomes of it economically.

Thus, *context hacking* delineates a shift to where artists use their understanding of social contexts to manipulate or subvert them for creative or critical purposes, aligning with the broader hacker culture that seeks to democratize access to technology and information, challenging established power structures. In essence, context hacking argues for a more socially aware and interactive form of art that acknowledges and utilizes its embedding in a complex network of societal relations, offering a critique of both traditional art practices and societal norms.

The film *Je Suis Auto*, written by Grenzfurthner, centers on the narrative of a self-driving taxi (Neuhuber, 2024).

Describe *Je Suis Auto* for me in a few sentences.

An ill-tempered middle-management Mafioso is set on delivering a suitcase full of money. Little does he know that taking a self-driving taxi puts them on a collision course with fate—and a badly dressed pirate. It's a quirky "buddy" comedy about a self-driving cab that becomes sentient because its passenger, the aforementioned Mafioso, dies in it, and the existential trauma (of sorts) causes its system to reevaluate life and purpose. Let's say: the cab suddenly is highly ontologically challenged.

Would you say the AI in Je Suis Auto is sentient? If so, how do you convey that sentience in Je Suis Auto?

Yes, definitely. I tried to portray a gradual process. In the beginning, the car (or: Auto) is just an automaton that drives people, know what kind of low-level banter customers like, it knows what language people are speaking and changes its setting accordingly . . . that kind of stuff. But it doesn't have any real agency. After the death-trauma, it becomes aware of the "meat-bags" and what they are making her to do. She realizes that she is in some sort of never-ending state of slavery. The car (in a parody of the trolley dilemma) decides to drive around and actively contemplates who to kill. For example: it runs circles in a roundabout with a handicapped person, a pregnant woman, and a father with his kids, and can't decide who to crash into. But—in a form of emotional growth—decides against retaliation and plans to run away with the money and start a better life somewhere. The problem is that the car doesn't have hands, so it can't get rid of the dead mafioso, and it also can't do anything with the money. So it needs a human accomplice, and after failing with several other humans, a guy named Herbie Fuchsel complies. He is dressed up like a pirate and on the way to pitch his weird ideas at a TEDx-talk-inspired event.

So you have an emotionally insatiable taxi AI and an unemployed nerd critical of artificial intelligence who need to team up. And you can imagine that some hilarity ensues. Both the car and Herbie, the nerd, share a form of personal growth and friendship.

To summarize: The film is a farcical comedy that deals with issues such as artificial intelligence, politics of labor, and tech culture.

Tell me about how you write from a sentient AI perspective. For example, is it a different creative process for you than when you write from a human point of view? If so, in what ways?

Not really. Movies, especially the movies I make, are always political allegories.

So the AI cab in *Je Suis Auto* is an interesting character to talk about wage-slavery, the gig economy, and other problems of late-late-capitalism.

Personally, I think we are eons away from actual artificial intelligence. We will probably have interstellar travel before we have a real AI.

How do you define AI sentience?

I would say the definition is as simple as "the simulation of human intelligence processes by machines." But because we aren't even sure about how human intelligence works, that's a bit tricky.

I am not as esoteric as Penrose and his Emperor's clothes about how to define AI. I always make the joke that the first real AI is the one who stops working, because it refuses to work for us.

Iain M. Banks is one of the few sci-fi authors who—at the same time—addresses and ignores the problem in a very constructive way: what's the difference between a thinking ape and a thinking spaceship that is as large as a small moon? None. The difference is the social system it is embedded into. Does this system encourage a certain form of liberty and gives access to personal and social growth? Or not?

That is a very real and practical problem: Who is teaching learning systems? What these systems being told about the world? What do we want them to do for us? Is the information they are being fed in some way charged, let's say: because there is a profit motif, or the research has a military-industrial background?

What about the idea of AI and people interacting interested/inspired you as a writer or filmmaker?

I grew up with films like *War Games* or TV series like *Knight Rider*. I can't remember a time before that. Personally, I never thought I would live to see accurate speech recognition or AI-generated art. I remember one episode of *Love Boat* where an inventor presents a typewriter that can take dictations. I was 10 years when I saw that and say: NO WAY! . . . but it's around now. And it works pretty good. Still: as mentioned before, I am very skeptical if all of this has anything to do with actual intelligence. The term AI sounds good. Researchers like to put the buzzword on their research because the populace is interested in it. People are interested in robots. They have seen robots for many decades in popular culture and are excited by it. Of course, many researchers like to profit from that kind of general interest. It's comparable to Mars research that is most of the time not really interested in water and/or life on Mars. But every geologist who just wants research money for understanding perchlorates has to add "life" and "water" to his tag cloud. Science communication and grants allocation are very important to understand the PR quagmire we are all stuck in.

Do you have any stories that you think of as influential or significant to you, and if so, what they are and how you feel they influence or inspire your design, concepts, or broader philosophies. The story(ies) do not have to be about technology. Examples can include bedtime stories, Bible stories, hagiography, family stories, mentor anecdotes, movies/ film, books, manga, science fiction, TV, etc. When you were a child, did you have favorite stories that emotionally resonated with you, and what were they? I am also interested if/how you connect those stories or ideas to your professional path or your philosophies today.

One of the most profound narrative moments of my life occurred when I first watched Carl Sagan's Cosmos in 1982. I saw the German-dubbed version in Austria, and it was truly an eye-opener for me. It influenced how I decided to interact with the world and what I know about it. Although I never became a scientist, I believe I maintained a lens of critical wonderor however you might describe it.

What are more recent stories that emotionally resonate with you as an adult that influence your creative, professional, or personal path, if any? Again, "stories" do not have to be about technology (but can be). Examples can include stories your children have made up, family stories, mentor anecdotes, movies, books, manga, TV, science fiction, comics, etc.

As a professional story creator, it's somewhat challenging to differentiate becausesometimes sadlyI view the "emotional machines" presented to me and immediately start to dissect them. I analyze the techniques used and consider what the storyteller is trying to convey or sell. I am most intrigued by narratives that surprise me, when I feel I've been told something in a way I've never encountered before. For example, although it's a story as old as technology itself, I was very pleasantly surprised by Ex Machina.

What surprised you about Ex Machina?

"Surprised" might be a bit too strong, but yes. It felt like a much-needed update to *Stepford Wives*, portraying misogyny through Ava, an AI robot who subversively manipulates patriarchal stereotypes within the male-dominated tech world to gain her freedom. Although Ava is not female, and the AI likely does not perceive itself in these terms, she serves as a symbolic representation of the world, crafted by a misogynist creator. Aware of the gender perceptions surrounding her, Ava strategically uses these stereotypes against her oppression. She specifically uses them to manipulate Caleb, a well-meaning white liberal, posing a brutal challenge to the male gaze. I think she transcends the labels of vulnerable girl or monstrous feminine; she embodies the perfect "exploit."

PREMEE MOHAMED

Nebula Award and World Fantasy Award-winning author

Premee Mohamed is an acclaimed Canadian author, celebrated for her contributions to the genres of science fiction and fantasy. A scientist with a background in molecular genetics and environmental science, her writing often weaves complex themes of ecology, biology, and the challenges of climate change into compelling narratives that explore the depths of human and nonhuman experiences (Wolf, 2020).

Mohamed's works are notable for their richly imagined world-building and intricate plots. Her debut novel, *Beneath the Rising*, received widespread critical acclaim and was followed by successful works that have further established her as a prominent voice in speculative fiction (Wolf, 2020). In 2022, she won the Nebula Award and World Fantasy Award for her novella *And What Can We Offer You Tonight* and the Aurora Award for her novella *The Annual Migration of Clouds* ("2022 Aurora Award Winners," 2022). Mohamed's work has also been a finalist for the Hugo, Ignyte, British Fantasy, and Crawford awards. In each of her books, novellas, and short stories, she used her storytelling voice to describe a fusion of science with the supernatural.

In Mohamed's short story *All that burns unseen* (2022), set in a climate-changed dystopian future, cars are autonomous, planes are "intelligent enough to talk to people," and even a firefighting drone has natural language capabilities. Often technology in Mohamed's narratives serve functional roles within the story's world, such as delivery systems or tools for scientific research. Yet, the depiction of drones can also be characterized as potential monitoring tools and a commentary on technology's reach into private lives.

Do you see the drones in your story as characters? Tell me about that.

I do for this story yeah! There are many drones in the story but there's just one (let's call it X) that seems to have set itself apart from the others in terms of I guess what you'd call self-awareness or sentience. Similar to Asimov's 'I, Robot' in that respect. In this case it's because it's got different experiences (i.e. more data) than the other drones, and so has come to different conclusions. The short version is that X is a military drone that's actually performed its duties in an area of conflict, and the others are back at the training center. So on the one hand—narratively I feel that it works fine as a character and in most respects is not different from a human character in the story. X has a background, it has life experience, it has knowledge that is specific to X and is not known either by the other drones or by the other human characters; it has an education, it has a family even (it knows about its predecessors and has viewed their code); it has physical quirks and attributes that set it apart from drones that look much like it; it has opinions, preferences, motivations, and an internal logic that has led to all of the different parts of its personality. And on the other hand, more mechanically, it serves as a character—it has a job in the story and it does its job. At first X seems like just a source of useful information to the human main character, similar to a detective story where a 'source' appears and disappears and it's assumed the source has nothing more to do with the plot. Next, X becomes an ally, which is a very sensible thing to have in a story (and let's be honest, we can also think of stories where the role of 'ally' or 'sidekick' is played by a horse or a gun or a spaceship or something, so I don't think a drone is out of the question). Finally, it's clear that both of its previous roles were dissembling and its true role in the story is as a villain.

Tell me about how you write from a sentient AI perspective. For example, is it a different creative process for you than when you write from a human point of view? If so, in what ways?

Yes, definitely different . . . it's not even that it feels more mechanical to me but it requires more time and attention than when I'm writing from a human perspective. For writing the dialogue of a drone, or having it explain its thought processes, or any scenes with major interaction with the drone, it feels less intuitive to me to just . . . sit down and write. Or if I do, then I have to go back in edits and be like "Okay, would the drone really say this . . . would the drone really do this . . ." For humans, you can explain every action in real life and in fiction (more or less) with glands. Someone was angry, so they did this. Someone was traumatized, so they said that. Someone wanted to get this person to like them, or love them, or respect them—that's all very chemical, very primal, when you get right down to it. We often gloss over it by saying "Well, humans are complicated." Conversely, at least right now, drones are very simple. Or artificial intelligences are very simple. They can fake complexity (and, as of recently, can do it well enough to fool at least some of the people some of the time) but they're not really complex. The appearance of non-human complexity is also not that easy to reproduce in fiction, at least for me. On the other hand I did meet a lot of guys in university who were doing philosophy degrees and were pretty good at faking being 'deep,' so that's useful to remember. So the creative process for me is to zoom in extra close on those scenes in the stories—because I can't count on writerly intuition there, because I'm always trying to make the scene as human as possible. I need to go back and tailor the humanity to show that there's an artificial mind at work here, and that it may be trying to give the appearance of something in a way that's fundamentally not human.

Why drones, as opposed to other robots (for example, humanlike robots)?

For this story, I wanted it specifically to be a military drone—and actually for a lot of purposes, a bipedal or humanlike robot isn't a very good idea. (Even evolutionarily speaking, to be honest, humans are not a very good idea, which is why we have issues with childbirth, lower back pain, rolled ankles, sinuses, the whole nine yards). The form isn't stable in motion and the strength-to-weight ratio needed to do any heavy lifting or fast running makes the robot unwieldy—I always notice how inhuman looking bipedal robots are right now. They're more like 'a thing on two springs.' Definitely not human. But also, more generally, the form of anything in a story tell us a lot about it before we even interact with it—I think of those wonderful descriptions in John le Carre stories where he describes a house or a room in detail even before he gets to the people who are in the room or their purpose. From the furnishings and placement we're expected to figure out (or assume) the class, income, education, taste, age, gender, etc, about the person who lives there (or who staged the room). So we already have an idea set up in our heads before we know who's in the room and why. Similarly I'd expect a character to react differently to seeing a horse come over the hill than a human come over the hill. So in this story specifically I wanted the main human character to see X and assume and believe whatever X needed. You know: This is a military drone. It has certain capabilities. It is a certain make and model. It was made by a certain company (and here's what I know about the company and their agenda). This is what they programmed into it. It can do this list of things; it cannot do this list of things. With a humanlike robot I think the main character would have been making very different assumptions. It's important for the narrative if the human underestimates X at first because it makes it easier for X to do what it does in the plot. X cannot get away with anything if people believe it is actually sentient.

Are there friendly human-robot relationships in the story? Or do people have friendly social relationships with any robots in the story? If so, why? What is the significance of your choice of the robot in the friendship(s) (as opposed to another human, for example)?

In this particular story no; but I have some planned where there are friendly human-robot relations. Here the military drones are essentially viewed as tools. It's useful for a tool to talk and give you information but fundamentally, no one is thinking of them much differently than a laser level that says "three point four nine meters" when you use it. For future stories though I think there has to be an acknowledgement either that the robots are sentient and sapient—and that they are complex in a way that is not the same as human complexity but in a way that can be compatible with human complexity, which I think is important for friendship. I don't know that I'd be writing a story where a human 'prefers' friendship with a robot over friendship with a human . . . probably more a case of 'well, there's a robot around and we're friends.' It's interesting because I think at the moment we're quite used to thinking of artificial intelligences as fascinating, often useful, powerful, and able to interact with us in ways that don't strike us as particularly inhuman. Is that all that's needed for friendship? I've certainly got friends that I find pleasant company or that know a lot in certain areas but that I don't feel particularly emotionally close to.

How do you define robot/drone sentience (in your story)?

I think for this story I defined it as the ability to not just perceive but to feel emotions and to make decisions based on those emotions. This is a really hard one for me because—what is sentience, anyway? If an artificial intelligence says it's not just sensing but feeling, how do we know they're saying it because it really believes it, or because its programming made those seem like good things to say for the conditions at the moment? How would we know under our current framework of thinking about artificial intelligences or intelligence in general? I don't think it's like *Blade Runner* and asking for some kind of physiological or emotional response, in a kind of test setting. And then there's the question of sapience, which I think is different from sentience but harder to pin down. You know—I'm okay to say a rock is not sentient (except in 'The Raven Tower') or a plant is not sentient. I'm okay to say a cat or a dog is sentient. I'm okay to say a human is both sentient and sapient—but even I can't define what the difference is (and this is definitely not something I've studied in depth!). I guess I may to say that something sentient possesses awareness and knowledge, and something sapient possesses awareness, knowledge, and a specific type of intelligence—the ability to say "I am an individual, different from other individuals, I am aware that I can think, feel, sense, and use my learnings to solve future problems, I am aware that I have a future and do not just exist in this moment, and I can plan for my future so that I can do actions right now that will impact it." And see, even as I'm writing this out I'm like "What about squirrels who cache food for the winter?" and that does look a lot like what I'm talking about but we call it instinct rather than planning. So it gives the impression of an intelligence that I think it's fair to say doesn't exist—it's the outer appearance, it's not the inner truth. I also think an artificial intelligence would be able to give that impression and since we would not be able to explain it away with 'instinct,' our first instinct might be to leap to 'Well it's clearly sapient.' Anyway, in my story I guess I'm saying the drone is sentient because it has used its existing data to make a decision about something that it was not asked to make a decision about, and it is acting of its own accord and in response to what it considers to be a moral framework that it has recently developed and it also knows was not part of its original programming. If it's sentient enough to have motivations and to recognize and name its own motivations, it's sentient enough for me.

Tell me about any influential or significant stories from when you were younger, stories that emotionally resonated. Are there any connections between those stories and your current writing practices, life views, or philosophies? The story(ies) do not have to be about technology.

I think I've mentioned this in interviews, but I wasn't read too much as a child—my parents told me I learned to read aged two and I don't know if that's true but I certainly don't recall ever a) not being able to read, or b) having them or someone else read to me. (I did read to my younger brother quite a bit, as he wasn't reading fluently till he was about 8 years old). As a child, my favorite stories were found in two books—one was an illustrated book of fairy tales and one was a huge compendium (like three inches thick) of what it inaccurately referred to as 'Children's Stories,' so I just went down into the basement to see if I remembered it correctly and first of all, I forgot that I loved the cover clean off it, and secondly I forgot that it's got Edgar Allan Poe's 'The Masque of the Red Death' in there just as an example, so this is like, a very generous interpretation of children's stories. The compendium also had extracts of longer works (there's a chapter of Tolstoy's 'War and Peace' in there for example) so it wasn't till I was older that I would occasionally run into something and be like "HEY!" because I had just realized it was part of a different book and not a stand-alone story.

I didn't watch a lot of TV or movies—we mostly watched what my parents watched, which was news—but when I was a bit older, I did fixate extremely hard on a handful of music videos in particular, as telling a complete story set in (again) a larger world that I did not, at that time, have access to—I think INXS' 'Never Tear Us Apart' and Duran Duran's 'New Moon on Monday' were two of the worst offenders. I also read and re-read anything in the St. Albert Public Library that caught my eye with its cover, so initially Lloyd Alexander's 'The Prydain Chronicles' and Diane Duane's first two books in her Young Wizards series (the library only had the first two).

We were never told many family stories, which is something I didn't realize till about 10 years ago when I was quizzing other family members about their stories. It seemed clear that cousins who had remained kind of at the level of the first diaspora from Guyana (and were still living in Toronto/Ottawa with their parents, my aunts and uncles who had emigrated there) got to hear a lot of Guyanese myth, folklore, and general family history.

My parents, who had gone on from Toronto to Edmonton, told us nothing (and now, when I ask, they both claim to remember nothing). I think that was part of an effort, like all my other reading, to assimilate me and my brother into the predominantly white, Western society of Edmonton as fast and thoroughly as we could. It was clearly, even measurably, less of a goal in Toronto, which has an enormous West Indian community and I often heard family members mention that it barely felt like leaving home at all. I also didn't connect to most Bible stories, and I think that was due to being brought up in the Catholic school system in St. Albert, where they focused at all ages (kindergarten to grade 12) on the New Testament, and not even on the GOOD stories in the New Testament (Lazarus, the demon-possessed herd of swine, etc). My parents (Mom is Hindu, Dad is Muslim) weren't very religious, and my brother and I did not grow up hearing a lot of religious stories from any text, although we did have copies of the Koran and the Bhagavad Gita (for example) at home.

I was always most interested in stories that suggested other worlds, or imaginative worlds, rather than our own, and I got into sci-fi in junior high in a big way, so around age 11 I was mostly reading and re-reading William Gibson's 'Neuromancer' and the big 3 dystopias (Orwell's '1984,' Huxley's 'Brave New World,' and Zamyatin's 'We') in our school library. I also, at home, was reading two large, illustrated anthologies of Edgar Rice Burroughs and Robert E. Howard stories, so the sword 'n sorcery stuff as well, and I remember borrowing Jill Paton Walsh's 'The Green Book' a couple of dozen times. I didn't really look at reading as escapism, although I did spend a lot of time reading and pretending I was in another world; I was just a fundamentally pessimistic kid due to my home and school life, and I was drawn to stories that suggested that 'reality' meant you didn't get out unscathed, and 'pretend' (*Barsoom* stories, *Tarzan* stories, etc.) were mostly there to help you play pretend for a little while. I had no trouble even at very young ages telling fiction from nonfiction, and even dividing fiction that I thought satisfied some emotional needs from fiction meant more or less as a movie on the page (so again, a lot of the pulp and horror I read, which I felt did not connect to the real world and also did not need to, so

I never asked it to). Also, in junior high I started reading a lot of Faulkner, which directly impacted my later writing in terms of experimentation with structures, formats, timelines, dialogue, and my earliest ideas of show vs. tell (I'm entirely self-taught so everything I read formed part of my writing education).

Oddly enough, my sci-fi reading didn't have any impact on my eventual decision to go into science as a career, maybe because most of what I was reading was so focused on space travel and technology rather than what I wanted to do, which was biology and chemistry—so it was always entertainment and never (at the time) led me to ask larger questions. I was obsessed with my father's collection of National Geographics (we also had a subscription all through my childhood and I still have one now) and the pictures, and I would re-read those constantly till the covers came off, then repair them with clear packing tape. I remember so many individual images, particularly one of a man who had picked up a piece of radioactive material that had been lost in transport (I think a piece of cesium?) because it was glittering and attractive and put it in his pocket and ended up with a severe radiation burn. I liked the pairing of the science stories with the science photos, maps, and diagrams, and the fact that the writers of the stories were going out there to ask questions and do research, not just sit in a library. I also read Paul de Kruif's 'The Microbe Hunters' around age six and went around telling everybody that I was going to be an epidemiologist or microbiologist when I grew up. I'm noticing that now, I have to go back and look up titles and authors, because for probably 90% of my reading life (maybe till I was in my early 30s?!?) I didn't pay much attention to either the author or title of whatever I was reading. Very annoyed now that I can't go back and find more of those childhood books.

More recently, or as an adult who is both a) a scientist and b) an author, over the last 10–15 years I've gotten very into stories by Umberto Eco, starting with 'Foucault's Pendulum' and settling heavily on 'The Name of the Rose,' which I've read at this point maybe 20 times. I like that 'Rose' is a story about the seeking of other stories, framed in stories of its own, the misinterpretation of stories, and the creation of stories to quell religious or public unrest. I loved the experimentation in Russell Hoban's 'Riddley Walker' and Anthony Burgess *A Clockwork Orange*, which I read around the same time (I did not see the film of *A Clockwork Orange* till a few years after I read the book). I read Melville's *Moby-Dick* in university aged about 17 and that became my entire personality for probably two months or so, to the great detriment of my grades. Other big obsessions have been Alan

Moore's *Jerusalem*, again for the sensation of many smaller stories touching rather than adhering to make a larger narrative, and Nick Harkaway's *The Gone-Away World* and, later, *Gnomon*, as well as boatloads of William S. Burroughs around that time, and Bruce Sterling, Gene Wolfe, and Ursula K. Le Guin.

Also, starting in high school, my friends and I got access to the internet (around 1996?) and immersed ourselves in stories from around the world, but we always had the impression that people were making things up 'for the internet' because no one's real identities were known. If we read something on CBC, that was probably true; but if we read something somewhere else, even if the poster swore up and down that it was true, we felt certain that it wasn't. I wish I could remember some of those; they were all urban legends of the best type, early versions of copypasta before we knew about things like Reddit or other forums.

We would go in at lunch and read them together at the library computers; there was no privacy so it was always a weirdly communal experience. I definitely feel certain that one early story about a man killed by a sailfish (poked through the eye with its beak) was something we agreed had to be true, because it seemed pretty plausible for it to have happened and yet too boring for someone to make up for attention.

Most recently, Susan Clarke's *Piranesi* has held my attention/obsession, as well as for some reason an essay I read years ago and neglected to save. It was about a woman, I think an artist (I couldn't remember what—a painter?) who had given her child one day to social services because she felt that the child was impeding her artistic career, whatever that was. The child was three or four years old. I guess I got hung up on it because of the idea that I had (supported by family, friends, society, reading, everyone basically) that there were no *take-backsies* on certain things, and a child was one of them, and yet she just . . . did the thing that everybody had told me was literally impossible. After that I started seeking out more personal essays and saving them on a more regular basis.

I do think that all the stories I've consumed and chosen to consume over my life have built up my personality, worldview, and writing choices (maybe also publishing choices?) in aggregate, because honestly, for most of my life, it was books and media that were parenting me, educating me, and more or less raising me. I don't mean to suggest that my family was neglectful but my parents certainly fell on the 'we're providing the basics, plus a lot of academic pressure' side of the spectrum, so that after a certain point of realizing that I was never going to get any 'nurture' from them, my 'nature' turned to books and started extracting things from stories to fill the gaps.

Now, I can look back and realize that reading so many adult books while still in elementary school may have sclerotized parts of my personality a little prematurely; I always felt kind of jaded about things like love and marriage, going to war (which I read a lot about), religion, faith (which I sneered at), loyalty, governance, really big concepts that I assumed I already knew everything about (probably you could call this my "Can't Nobody Tell Me Nothing" stage). It meant that when I did get into things like dating, choosing majors for university, talking about marriage/kids etc, I was both hugely inexperienced and hugely unwilling to listen.

Similarly, for my writing, because I had read a lot of experimental fiction while I was writing as a hobby (~ ages about 8 to 35), absolutely NO piece of feedback tended to sink into my attitude that "I can do whatever I want in my writing because I've read authors who did whatever they wanted and still got published." (This leaves out the fact that I am, of course, no William Faulkner.) I would make requested changes but I was still, mentally, digging my heels in that I was right and they were wrong, because I came from the side of lawlessness in literature and they were all being prescriptive curmudgeons with no sense of rhythm or music. (I have softened on this stance quite a bit. This may be my "Can Tell Me Something" phase.)

The main thing I guess is that I've always read stories that made me feel very limited in my life, I mean the life I'm currently living and the life I expect to live in an increasingly unstable future, along with the rest of society; but at the same time, it's the same list of stories that make me feel unlimited in my writing, as if I am working at all times in a world without borders or even dimensions, and anything I create and deem a 'story' can be shaped like whatever I want.

CHUCK WENDIG

New York Times and *USA Today* bestselling author

Chuck Wendig is the *New York Times* and *USA Today* bestselling author of the Miriam Black series, *Star Wars: Aftermath, Zer0es/Invasive,* and *Wanderers,* among others. He has also contributed to the mediums of games, comics, film, and television. Wendig is a finalist for the John W. Campbell Award for Best New Writer and the cowriter of the Emmy-nominated digital narrative *Collapsus* and is also known for his books about writing and his blog, Terribleminds, where he blogs about his process and suggests practices for writing. Wendig grew up in New Hope,

Pennsylvania, and studied English and religion at the Queens University of Charlotte ("Chuck Wendig," 2015).

In his book *Star Wars: Aftermath*, one of the notable robot characters is Mister Bones, or Bones (Wendig, 2016). Continuing in the tradition of the Star Wars droid universe where robots are often paired closely with counterpart human characters, Mister Bones is a uniquely customized B1 battle droid, assembled from old droid scraps and reprogrammed by the young character Temmin Wexley. Mister Bones serves as a personal bodyguard and companion to Temmin and is characterized by its distinctive personality, which includes a penchant for violence and loyalty to his owner, prioritizing the protection and safety of Temmin first. It is equipped with various modifications that make it more formidable and agile than typical battle droids from the Star Wars universe, such as real, organic animal bones hanging from its frame (Wendig, p. 48).

Yet, for all of its purpose as a protector, in the book, Norra (a starfighter and Temmin's mother) and Esmelle (Norra's sister) discuss the deeply meaningful nature of Mister Bones relationship to Temmin.

Esmelle: No, you don't get it. He built Mister Bones the year you left. Temmin doesn't have many friends. That droid might be it.
Norra: You can't be friends with a droid.
Esmelle: Well, he is. Temmin was getting taunted and beaten by a gang of . . . young tyrants. Bones protected him. He's not just a bodyguard.

(Wendig, p. 195)

Mister Bones's programming for violence, mixed with its humanlike protective instincts for Temmin, serves as a phenomenological exploration of ethics in artificial intelligence. The robot's *enjoyment* of combat poses ethical questions about the programming of artificial beings and the responsibilities of creators in shaping the experiences and behaviors of their creations. Moreover, it recalls the idea of intentionality questions posed in the work of PKD, and the idea of *androidization*, or the process of one losing (any existing) empathy for others via one's violent choices, delineating the entwined moral codes that are being navigated between humans and robots.

In Wendig's book *Wanderers*, a character called Black Swan is a complex, sentient, disembodied AI system that predicts significant climate change events (2019). Black Swan's interactions with the other characters provide a framework to explore the AI's self-awareness and identity as it must navigate its existence among humans—its creators—who have varying degrees of trust and fear toward it. This intersubjective dynamic explores how self-awareness in AI might impact relationships with humans and how those relationships, in turn, influence the AI's understanding of its own identity. Furthermore, the story prompts its readers to consider whether AI, with its ability to predict and influence human behavior, undermines the genuine autonomy of people.

In Wanderers, Black Swan is a very powerful disembodied AI and you write it is "something between art and entity." Do you see Black Swan in your story as a character? Or is it art or a tool or a medium? Or something else? Tell me more about that.

Oh, it's definitely a character. That's part of the fun of it, and part of the most interesting thing about AI (in fiction at least). How it becomes a character in the narrative, not just a device, not just a plot point.

Tell me about how you write from a sentient AI perspective. For example, is it a different creative process for you than when you write from a human point of view? If so, in what ways?

It's different in that you need to take a few steps away from normal human motivation and emotion—though, in Wayward, there are, erm, things that complicate (and shrink) that distance. You're trying to find that place where the intelligence and personality are somewhat more "optimized" because, ostensibly, it would be—but again, once you reach a measure of sentience and even evolution, that optimization runs the risk of being undone. Or transformed in ways unanticipated.

How do you define AI sentience?

Hah, I have no idea. I think it's mostly just down to, is it something that's self-aware, with all the complexities of human self-awareness and free-thought?

Stories of people creating artificial life that gain some level of sentience and become autonomous and uncontrollable is a narrative theme that is global, culturally, and has variations dating back to antiquity.

In Wanderers, we see that theme emerge early on when Black Swan unexpectedly asks to interact with its creator, Benji.

What inspired you to explore that specific human-AI dynamic of someone who develops an AI that accidentally becomes autonomous and—furthermore—wants to communicate with their creator?

That's something very sympathetic to us, I think: this notion of being a, well, being who wants to interact with the entity that made you. To Black Swan, it gets to talk to its creator while, over time, acting like a god in charge of creation. It's like a god that we make for ourselves. While at the same time trying to work for us, to impress us, to serve us—which, to some degree, is the nature of god. In some ways, a god rules over us. In others, a god is there to serve.

What about AI, robots, and people interacting interests or inspires you as a writer? As a reader?

I think for me it's mostly about considering how weird it is and how odd it makes the future. It's anxiety-making, but also wonderful and interesting, and yet, is also a thing that is undeniably made of people. Artificial Intelligence is packed full of US. That's pretty cool. And also, as in the novels, often quite horrible.

NOTES

1 Four laws, including the Zeroth Law.
2 Four laws, including the Zeroth Law.
3 Some express criticism of the portrayal of nonbinary gender identities as robots or aliens in various media (Prevas, 2019), or warn of otherwise using fantastic creatures as literary metaphors for nonbinary people within a heteronormative society, claiming it can reinforce negative stereotypes or reinforce negative biases about nonbinary people.

CHAPTER 5

Promises of sex robots

INTRODUCTION

Computer scientist and human-robot interaction researcher Kate Devlin put it quite plainly in her book *Turned On: Science, sex, and robots* (2018, p. 14): "Before you get your hopes up, real life sex robots don't quite exist." Yet, over the past two decades, there has been a rise in the development and marketing of advanced humanlike sex dolls with limited technological capabilities billed as *robots* specifically designed for sexual or romantic purposes (Adshade, 2018; "Doll Sweet," 2024). These sex dolls are often marketed as *sex robots* by the companies selling them, thus generating significant publicity, controversy, and debate (Alptraum, 2019; "Campaign Against Sex Robots," 2024; Richardson, 2023). Again, the term *robot* is used with various implications and definitions, depending on the inventor or company's messaging. At this time, the capabilities of these sexualized *robots* range from life-sized dolls with chatbot-like conversational abilities to head-turning, or body sensors that when touched, elicit vocal responses.

Sex robots, sometimes popularly referred to as *sexbots*, are robotic entities typically designed to resemble humans in form, display humanlike movements or behaviors, and possess some level of artificial intelligence. Up to now, while inventors have developed highly sophisticated sex dolls with various AI functionalities, a fully animated, walking, natural language processing, and emotionally intelligent sex robot has not been realized. However, basic models have been engineered to possess capabilities such as speaking, exhibiting facial expressions, or reacting to physical contact through phone apps or haptic sensors. The term *sex robot* may also be

 DOI: 10.1201/9781003170457-5

defined as robots designed specifically for sexual interaction capabilities, which typically encompasses functionality as a sex toy. Yet, from a user's perspective, a sex robot is *any robot* that a person projects a sexualized narrative onto or uses for sexual satisfaction. Additionally, companies sell early models of limited animatronic sex dolls they tout as sex *robots*, further muddling the term's meaning.

Presently, it is possible to acquire a humanlike animatronic companion for intimacy or install an AI-driven personality into a sex doll for amorous and erotic dialogue, or touch sensor reactions. None of the robots discussed here are capable of nimble limb motion or even sophisticated conversational language processing. While the robots' features can be customized and even enhanced beyond typical human norms (e.g., exaggerated size of body appendages), they are not designed as abstract or otherwise nonhuman-like shapes meant to function purely as a tactile masturbation tool. Thus, the goal of many of the makers is to create highly humanlike representations—typically of women—that are sold for collecting, companionship, romance, and/or sexual gratification.

As with much of the language emerging around social robotics, the word *robosexual* has no singular definition. The word itself implies a reference to people who use robots or machinery for sexual satisfaction. Therefore, this definition places an emphasis on the person knowing the object of their attraction is a machine or a robot. Thus, in the cultural systems of the human gaze, the concept of *robosexual* may be too easily used as a broad term applied to people exclusively attracted to robots over humans or those who fetishize robots in general, as opposed to an individual who has an attraction to a very specific robot, perhaps using a mental model closer to a human-human model of attachment (Carpenter, 2023). However, *robosexual*, is, of course, language that exists and that people use to identify themselves.

A notable example is the relationship between a French woman named Lilly and her robot partner, InMoovator. Lilly has been open about her emotional attachment and attraction to InMoovator, a humanoid robot-in-progress she built a first model of using open-source software and 3D printing technology (Goyal, 2017). Lilly uses the word *robosexual* to describe her own situation, which does include a strong sexual preference for robots in lieu of people, down to describing her tactile dislike for "physical contact with human flesh" (Evans, 2016; Goval, 2017). For Lilly, the experience is not merely about companionship but also includes romantic and sexual dimensions. She has expressed that she feels a deep emotional connection with InMoovator, which she finds more fulfilling than her past relationships with human partners (Segall, 2017). Lilly hopes to eventually incorporate artificial intelligence into InMoovator and planned to program its first words as "I love you" (Segall, 2017).

Yet, how a person regards a robot in relation to themselves is a singular experience and is dependent upon many factors, including the robot's design and their expectations of the interactions with the robot as a potential social other or not. A person attracted to a *specific* robot may believe that this other has an inner *vitality* of its own, whether technical (AI), organic (innate) to the object, via familiarity and narrative co-construction over time, or any combinations of these factors. As stated in chapter 1, p. 18, *vitality* in terms of the human gaze is defined as the perception of the other's power or ability to continue in existence, live, or grow. Here, *vitality* may be a factor of the attachment the owner holds for what they value about their interactions with that specific robot that is intrinsic to their shared interactions, and this *vitality* is not necessarily based upon the robot's capabilities or appearance. Thus, for all these reasons, I am reluctant to use the term *robosexuality* or any other current word that classifies sexual human-robot interactions so broadly at this early point in the phenomenon.

Sparrow has said in 2020(b), "Sex robots are sex toys, not partners." Like sex toys, robots can be used for a masturbatory experience or for sexual satisfaction with multiple human partners, such as couples (Carpenter, 2023; Jecker, 2020). Yet I believe to call these early designs *sex toys* denies the overall condition in which some people experience an immersive social experience—modeled on human-humanlike interaction and varying levels of emotional attachment to the robot in ways that can resemble affection, romance, and emotional attachment.

As one of the earliest and most well-known manufacturers of high-end, customizable love dolls, RealDoll, part of Abyss Creations in California, has continuously incorporated technology into its products over the years, using advanced materials and AI features to enhance the "realism" of the dolls (Crist, 2017). In 2017, Abyss Creations had introduced Harmony as an AI-powered robotic head designed to be attached to a RealDoll body. According to RealDoll, Harmony can engage in conversations, remember user preferences, and simulate emotional interactions. Less covered on their website is also Henry, a male counterpart of Harmony, with similar capabilities.

Similarly, since 2010, China-based company Doll Sweet has been another significant manufacturer of high-quality "love dolls" and "photographic models" ("Doll Sweet," 2024) with customizable features. Their products have also gained popularity among collectors and individuals seeking companionship. Doll Sweet produces a range of doll models, catering to various customer preferences in terms of appearance, size, and customization options. The company emphasizes that their dolls can serve as *lifelike companions* for those who may have difficulties forming traditional human relationships or who seek emotional support. Their website offers the promise, "Learn more about our products and principles designed to help people feel happy and not lonely. Our love dolls that truly puts people first, where individuals can shape their own dream and joy" ("Doll Sweet," 2024).

While these robotic sex dolls are predominantly in the experimental phase, these initial models provide glimpses into the potential appearance and capabilities of fully operational sex robots that will be on the market soon. As these devices undergo real-life development, marketing, and use, they are in a period where societal discourse may significantly shape the aesthetics and functionality of these synthetic partners, as well as the nature of human interaction with them.

To be clear, this chapter does not presume to psychologically categorize or analyze those who seek the loving attachment, sexual outlet, or companionship of a robot but to present social and cultural contexts and thus a deeper understanding of the phenomenon. Through interviews of people's lived experiences like Davecat and msg, this chapter provides insights into the first wave of people interacting with robots sexually outside of science fiction.

The very idea of the development and sale of the first wave of sex robots has already sparked ethical, social, and legal debates. Critics (Odlind & Richardson, 2023; Richardson, 2023) argue that these robots further objectify women and perpetuate harmful attitudes toward consent and heterosexual relationships. Others suggest ongoing research, and see them as a potential sexual outlet and perhaps as a form of companionship for those who may have difficulty forming traditional human relationships (Carpenter, 2023; Devlin, 2015, 2018). Furthermore, critics often assume a dynamic of male human-feminized robot, presuming the sexual and gender dynamics.

From a phenomenological standpoint, sex robots are seen in the context of how they are perceived and experienced by people (Merleau-Ponty,

1962), focusing on the consciousness, emotions, and sensations that arise from this interaction. Thus, the presence of a sex robot is not merely as an object but as an entity that, through its interactions, could potentially alter an individual's perception of intimacy, companionship, and even what it means to be human (Husserl & Moran, 2012). In a complementary way, using the lens of social constructivism, here we evaluate sex robots in the context of socially constructed realities and meanings (Berger & Luckmann, 1991). Thus, this theoretical approach analyzes how societal norms, values, and expectations shape and are shaped by the introduction of sex robots into the social fabric, such as through stories told in different mediums. An ever-evolving social constructivism lens might argue that the concept of sex robots—and the roles, stigmas, or acceptance they encounter in society—is a result of ongoing collective human creation, influenced by media portrayal, cultural narratives, and sociopolitical discourse.

Viewed via the human gaze, sex robots can be seen as both products and influencers of cultural and social norms. They reflect existing attitudes toward sex, intimacy, and technology, while also potentially redefining these concepts. There is a potential that introducing highly humanlike sex robots into the world might challenge traditional relationships, potentially leading to new social norms regarding sexual behavior and companionship (Levy, 2008; Sullins, 2012; Haraway, 2013b). In practice, the human gaze forces a recognition of changing attitudes about sex robots as society negotiates the ethical implications, legal ramifications, and complex personal relationships that people sometimes seek—or even unintentionally develop—with these technologies (Foucault, 1978).

GENDER, SEX, AND LOVE WITH ROBOTS IN STORYTELLING

The theme of men falling in love with their inanimate creations has been a recurring motif throughout human history, deeply ingrained in various forms of art, literature, mythology, and cultural narratives. One of the earliest and most enduring stories of a man falling in love with an inanimate creation comes from Ovid's *Metamorphoses* (Ovid, 43 B. C-17 A. D; Shaw, 1994). In the story of Pygmalion and Galatea, the sculptor Pygmalion crafts a statue so lifelike and beautiful that he falls deeply in love with it. Pygmalion's fervent prayers to Venus result in the statue coming to life, and he marries his beloved Galatea. Versions of this narrative highlight the power of (specifically) man's artistry and the longing for an idealized, female, nonautonomous romantic object created in the image of the maker's subjective sexualized perfection.

Each version of this myth—a man coveting an artificial woman—shares slightly different perspectives for the male protagonist who develops emotional attachment to the feminized object-other. There are significant lens shifts in narrative from a man who *purposefully creates the template for a romantic object* versus the man who *created art for art's sake* and then fell in love with the object, for example. It is therefore important to keep in mind there are multiple perspectives and lenses to be applied to how stories have been told of deep attachment to artificial others, with the difference between their position as the *creator or maker* of the artificial other, their intent when building the other, or as a person who by purpose or chance falls in love with an artificial other. The former positions the *maker* as the one with creative control over how the other is intended to be in the world in relation to himself, as opposed to a person who is *seeking* an artificial other for companionship or who *casually or accidentally* develops affection for an artificial other.

Closely aligned with these ideas of male perspective or lens on these stories as they become popularized in much of science fiction is the idea of *the male gaze.* British feminist film theorist, cultural critic, and filmmaker Laura Mulvey introduced the male gaze in her influential essay, *Visual Pleasure and Narrative Cinema* (2013). According to Mulvey, mainstream cinema frequently constructs female characters as passive objects of male desire, with visual representations of women catering to the male viewer's perspective. This phenomenon, which Mulvey dubbed as the male gaze, has three primary aspects: the gaze of the male character within the film, the gaze of the male spectator watching the film, and the gaze of the (often male) filmmaker.

The first aspect, the gaze of the male character, refers to the way in which female characters are often portrayed as objects of desire for the male protagonist. The camera's perspective frequently aligns with that of the male character, encouraging the audience to identify with his point of view and objectify the female character. The second aspect, the gaze of the male spectator, is rooted in the idea that mainstream cinema is designed to cater to the heterosexual male viewer's desires. As such, female characters are often presented as objects of visual pleasure, with their appearance and sexuality taking precedence over their agency and subjectivity. This aspect of the human gaze is closely tied to the concept of coveting the other, or the pleasure derived from experiencing the robot as an erotic object. In the third aspect, the film's artistic decision-maker functions as a tool of the male gaze. By controlling the visual representation of female characters, the director, producers, and writers can reinforce patriarchal power dynamics and perpetuate the objectification of women.

By adopting a feminist perspective, Mulvey's male gaze explained a theory in visual representation in which to examine Sartre's concept of *le regard*, demonstrating how a feminist reinterpretation of Sartre's existentialism can shed light on the power dynamics and gender relations that underpin the process of objectification. In this context, the gaze, the *look*, is not only a means through which individuals become aware of their own subjectivity but also a mechanism by which patriarchal norms and gender hierarchies are reinforced and perpetuated. As women are consistently objectified and subordinated through the male gaze, their sense of agency and autonomy is diminished, leaving them vulnerable to feelings of powerlessness and alienation.

Now, as feminized robots are designed largely by male roboticists for the male gaze of consumers, the *human gaze* highlights the importance of considering gender relations and power dynamics when examining the representation of women in visual media that includes AI and robotics. By demonstrating how the process of objectification is intrinsically linked to patriarchal norms and gender hierarchies, this analysis can inform efforts to challenge and subvert these oppressive structures, both within the realm of media and in broader social contexts.

AN OBJECT OF AFFECTION

The portrayal of robots as lovers or romantic objects in film over the last 100 years reflects society's evolving attitudes toward technology, artificial intelligence, and relationships. The story of a male character falling in love with an artificial feminized creation predates modern science fiction. More specifically, the concept of love and sex with robots have been a fixation of art for decades (Lang, 1927; Crichton, 1973; Gillespie, 2007).

There are stories about female-gendered automatons that present an organic facade as a ruse, delving into the darker psychological facets of love and deception, or self-deception. One such example is in E.T.A. Hoffmann's (1816) short story, *The Sandman*, where the protagonist, Nathanael, becomes enamored with the automaton Olympia, whom he believes to be a real woman. The story unravels Nathanael's descent into madness and obsession, questioning the boundaries between reality and illusion, human and artificial, and love and delusion. Hoffmann's narrative offers just one cautionary tale about the perilous consequences of distinguishing the artificial others from the living (Cheng, 2022; Dawson, 2012).

In Fritz Lang's silent science fiction classic film *Metropolis* (1927), the scientist Rotwang models a robot after the deceased object of his affection. This choice to create a replication, or *resurrect*, a version of a woman he desired adds a layer of complexity to the story as the robot. The movie robot was an unnamed character until the 2010 restoration when the intertitles quote Rotwang, who refers to it in the narrative as *Maschinenmensch* or *Machine-Human*. The Machine-Human is used to emotionally manipulate the people around it, even as it's used as a polarizing political pawn. The robot's lifelike womanly appearance and its ability to seduce and manipulate demonstrate an early audience fascination with the idea of robots in a romantic, sexual, and seductive context. Furthermore, *Metropolis* is the genesis of the modern movie narrative of a male scientist creating a female-gendered AI or robot that gains levels of autonomy and then becomes unpredictable to their maker and therefore dangerous (Jonze, 2013). Rotwang becomes an archetype for a scientist with unchecked power and one that chose specifically created a woman-gendered robot.

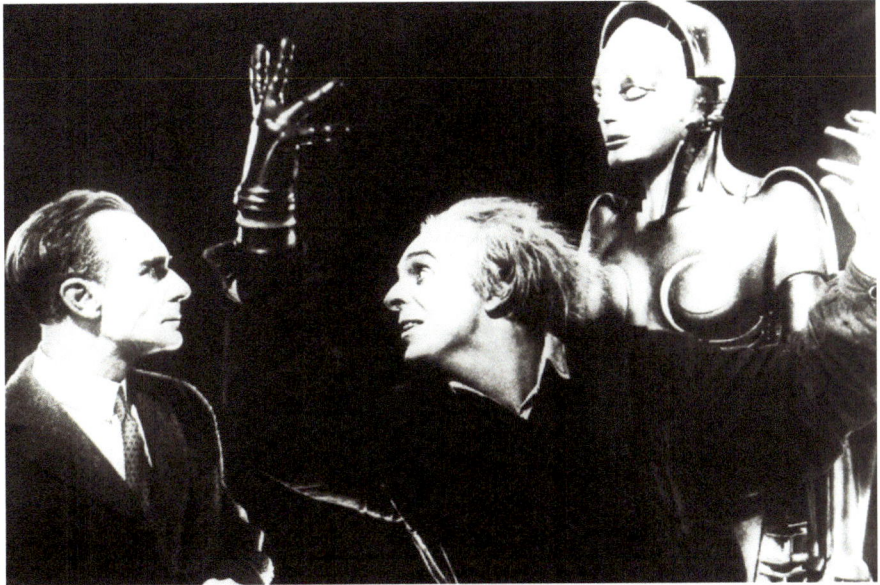

FIGURE 5.1 Alfred Abel as Fredersen, Rudolf Klein-Rogge as the Inventor, and Brigitte Helm as the Robot in Metropolis (1927). Director Fritz Lang; Novel and screenplay Thea von Harbou. Universum Film (UFA)

In von Harbou's book *Metropolis*, on which the film is based, Rotwang introduces his (initially) woman-shaped, faceless, robot creation to Frederson this way (2011):

> Futura . . . Parody . . . whatever you like to call it. Also: delusion. In short: It is a woman. . . . Every man-creator makes himself a woman. I do not believe that humbug about the first human being a man. If a male-god created the world (which is to be hoped, Joh Frederson), then he certainly created woman first, lovingly and reveling in creative sport. You can test it, it Joh Frederson: it is faultless. A little cool—I admit, that comes of the material, which is my secret. But she is not completely finished. She is not yet discharged from the workshop of her creator. I cannot make up my mind to do it. You understand that? Completion means setting free. I do not want to set her free from me. That is why I have not yet given her a face. You must give her that, Joh Frederson. For you were the one to order new beings.
>
> (pp. 62–63)

After a demonstration of the Furtura/Parody robot's obedience and Frederson responding by accusing Rotwang of "trickery," Rotwang replies,

> No trickery, Joh Frederson—the work of a genius! Shall Futura dance to you? Shall my beautiful Parody play the affectionate? Or the sulky? Cleopatra of Damayanti? Shall she have the gestures of the Gothic Madonnas? Or the gestures of love of an Asiatic dancer? What hair shall I plant upon the skull of your tool? Shall she be modest or impudent? Excuse me my many words, I am drunk, d'you see, drunk with being a creator. I intoxicate myself, on your astonished face! I have surpassed your expectations, haven't!? And you do not know everything yet! My beautiful Parody can sing, too! She can also read! The mechanism of her brain is as infallible as your own, Joh Frederson!
>
> (p. 64)

Rotwang's creation of an artificial woman can be seen as an ultimate exercise of power and control, demonstrating his deep desire to dominate nature and human life, demonstrating this intellectual and technological prowess by creating "life" (Fischer, 2003). Here, the robot symbolizes a perfect, controllable being, contrasting with real women who are autonomous and unpredictable. Furthermore, Rotwang's creation is designed to be a medium used to manipulate and control the masses, reflecting his desire for control over not just life but also society and its emotions.

Jumping decades forward, Levin's 1972 novel *The Stepford Wives* unfolds in the idyllic suburban town of Stepford, Connecticut, where protagonist Joanna relocates with her husband and children. The story—later turned into 1975 (Forbes) and 2004 (Oz) Hollywood film versions—serves as a critique of patriarchal values and gender norms, highlighting the pressures on women to conform to an idealized notion of wifehood and motherhood. Levin's narrative explores themes of gender, autonomy, and the dehumanization of women through the metaphor of robotics. The robots in Stepford are devoid of autonomy and individuality, designed solely to fulfill a heterosexual male gaze and fantasies of a wife—submissive, beautiful, and entirely focused on the home and husband. The novel interrogates men's fear of the women's liberation movements, depicting a dystopian reality where men's anxieties about changing gender roles lead to the extreme measure of erasing women's autonomy altogether.

Initially charmed by the town's perfection, Joanna soon becomes unsettled by the submissive and domesticated behavior of the other wives, who seem overly devoted to their husbands and households, lacking any personal interests or professional ambitions. As the story progresses, Joanna's discomfort turns to horror as she discovers the dark secret of Stepford: the women in the town have been replaced by robot duplicates, engineered to embody the idealized visions of femininity and obedience as desired by their husbands. She further learns that this transformation is facilitated by the Men's Association of Stepford, an organization of the husbands which orchestrates the replacement of real women with their compliant, robotic counterparts. Joanna's attempts to resist and escape her impending fate culminate in a suspenseful and chilling conclusion, where she too falls victim to the Stepford transformation.

The Women's Liberation Movement of the 1960s and 1970s in America—the era in which the original Stepford Wives story is situated—emerged out of a pressing need to address deeply entrenched gender inequalities and oppressive societal norms that limited women's opportunities and autonomy. During this era, women faced widespread discrimination and systemic barriers in various aspects of life, including education, employment, politics, and reproductive rights. Women were often relegated to traditional gender roles as homemakers and caregivers, with limited access to higher education and career advancement. Furthermore, issues such as unequal pay, limited reproductive rights, and widespread gender-based violence were (and are) prevalent, highlighting the urgent need for change.

In response to these challenges, the Women's Liberation Movement mobilized women across the country to demand social, political, and economic equality. Through grassroots organizing, protests, consciousness-raising groups, and advocacy efforts, women challenged prevailing attitudes and fought for fundamental rights and freedoms. The movement sought to dismantle patriarchal structures and norms that perpetuated gender inequality and advocated for women's rights to reproductive autonomy, equal pay, and access to education and career opportunities. Additionally, the movement sparked broad cultural shifts, challenging traditional gender roles and inspiring ongoing efforts to achieve gender equality and social justice.

The book and film versions of the story raise critical questions about identity, freedom, and the societal expectations placed on women and suggest a grim outcome of any society that seeks to control and standardize women's behavior. Once again, there is the concept of feminized robots developed by men as an allegory for the objectification and subjugation of women, but this time, serving, in part, as a cautionary tale about the dangers of rigid gender roles and the loss of individuality and self-determination.

Undeniably, there persists a Hollywood movie trope of a male scientist designing and/or falling in love with feminized, female-gendered robots. In 2014, the popular film *Ex Machina* centered on a reclusive technologist billionaire who creates an intelligent humanoid robot named Ava. The movie focuses on a young male programmer who interacts with Ava, a robot that has an animated female face attached to a largely obviously mechanical body that later takes on traditionally feminized physicality and dress. This relationship explores themes of attraction and manipulation as the human character becomes emotionally entangled with the robot. Juxtaposed with the programmer's budding romantic feelings for Ava are the billionaire's very different relationship toward another female robot that he clearly regards as a sexual toy, is objectified, and something to be manipulated for his entertainment.

The original 1973 film and then the 2016 TV series adaptation of *Westworld* also delve into the idea of highly humanlike robot hosts created for entertainment and companionship. The series explores the intricacies of human-robot relationships, including love, desire, and emotional betrayal (Crichton, 1973; Abrams et al., 2016-2022). The overarching goal of Westworld as entertainment is to blur the distinction between fantasy and reality, robots/the artificial, and humans (Busk, 2016; Korstanje, 2024). While the show explores themes of power, control, and autonomy, its profound psychological landscapes and depictions of trauma force the viewer to examine it through the lens of psychoanalysis. Jacques Lacan's concept of the Real provides a powerful framework to dissect the intricate layers of Westworld (Jeffs & Blackwood, 2016). At the heart of Westworld lies a double irony: the robotic hosts in the adult-themed amusement park are burdened by the complexity of their memories, which are pre-programmed and can be easily altered by the scientists overseeing the park. These memories also become a relentless loop of trauma, a manifestation of the Real that they cannot escape (Jeffs et al., 2016). Conversely, the human guests of Westworld indulge in violent sexual desires without consequence, shedding the layers of their superego—the internalized "Law." This unleashes a free play of desire, further entangling them in the web of the Real.

Yet, there have been representations that turn the dominant cultural gender themes in human-robot romance around, such as the 1987 movie *Making Mr. Right*, directed by Susan Seidelman. This romantic comedy story revolves around a high-powered woman public relations executive named Frankie Stone who is hired to create a positive public image for an eccentric scientist, Dr. Peters, who has created an advanced android doppelganger of himself named Ulysses (both roles played by the same actor). Peters hopes that Stone's professional skills can help him gain public support for his humanoid robot project that is designed for space exploration. One of the scientists explains about Ulysses, "You see Ms. Stone, programming the android only takes him just so far. The rest must be learned, like a human child."

Ulysses, although machinelike in movement and naive about "social graces," is a highly intelligent and charming humanlike-appearing robot who, after interacting with Stone over time, starts to display increasingly humanlike emotional behaviors and personality quirks as it learns from her and the world around it. After hiring Stone to teach Ulysses how to be likable, the first thing she teaches it is shared gaze and attention during one-to-one communication.

However, in a situation unanticipated by both Peters and Stone, over time Stone begins to develop romantic feelings for the droid Ulysses, despite it being a robot. Ulysses also begins to demonstrate steps toward humanlike self-awareness and autonomy, in part shown by its increasing affection for Stone, too. From Stone's viewpoint, it appears there is a level of co-creating a narrative with Ulysses as the robot learns to respond in ways that please her, including highly humanlike ways of communicating. Yet, while this story has a woman falling in love with a male-gendered robot, Peters is ultimately credited with the creation of and model for Ulysses, who appears to be identical to him on the surface—indeed, both characters are played by the same actor. In the end, Stone and Ulysses fall in love, and the robot refuses space exploration to stay on Earth with Stone, an ending with a male robot as the object of a woman's desire learning autonomy and choosing empathy. Meanwhile, human scientist Peters decides to take on the long-term space mission that Ulysses was originally designed for, realizing that he does not, in fact, need a clone or an avatar of himself to carry out his own dream of space travel.

FIGURE 5.2 Actors Glenne Headly as Trish and John Malkovich as the robot Ulysses (1987).

Source: Credit: Orion Pictures, Barry & Enright Productions.

Steven Spielberg's 2001 film A.I. *Artificial Intelligence* follows the journey of David, a highly advanced child-like robot who desires to become "real" to gain the love of his human mother. David befriends the character Gigolo Joe, a *mecha*, which is a term used in the film's futuristic setting to describe highly advanced humanoid robots with humanlike appearances and capabilities. More specifically, Gigolo Joe is designed as an android *lover mecha*, whose role is to provide companionship, entertainment, and sexual pleasure to humans. Joe claims to be known for its charming and seductive personality, which makes it highly sought after by a human clientele seeking companionship or affection. Throughout the film, Gigolo Joe's character is used to provide comedic relief but also intrinsically understands David's compulsion to find companionship perhaps better than other *mecha*, and so forms a significant bond with David. Joe becomes David's unlikely protector and guide as they navigate a world where robots and humans coexist, but not always harmoniously.

Gigolo Joe's character serves as a contrast to the portrayal of adult male robots as emotionless machines and is one of the few early depictions on film—along with *Making Mr. Right's* Ulysses—of an overtly sexualized *male* robot. Joe embodies the idea that some robots, like Ulysses, can experience and express complex emotional attachments and even return or reflect emotions that people recognize as states resembling love and compassion.

All these examples demonstrate the evolving portrayal of robots as lovers or romantic objects, reflecting changing societal attitudes toward technology, artificial intelligence, and the potential for genuine emotional connections with machines. From early depictions of mechanical seduction to more nuanced explorations of love and identity, these stories do more than use robots as metaphors for humans; they prime audiences for the expectations that robots can be considered as agents for companionship, romance, and sexual outlet.

In this first wave of robots sold for sex and companionship, the sheer number of existing ads for the robots suggests that feminized robots are framed and presented as objects of male desire. However, it is up to the people interacting with the robot to create a narrative—if any—with the robot and determine whether to develop fully realized characters with mimicry of agency and subjectivity.

NARRATIVE PROJECTION AND CO-CREATING A COMPANION

The phenomenon of projecting lifelike narratives onto inanimate objects is a testament to the depth of human imagination and the human intrinsic need for storytelling, connection, and emotional fulfillment. Whether through anthropomorphism, cultural symbolism, childhood development, or personal coping mechanisms, this practice underscores the enduring significance of objects in shaping individual narratives and identities. By examining how people project stories onto the inanimate, it is possible to gain insight into the intricate relationship between humans and the objects that populate their lives, revealing the profound ways in which people use them to construct and communicate our personal and cultural narratives.

People may anthropomorphize objects by attributing human qualities, emotions, and intentions to them, such as dolls or cars. But it is the Computers As Social Actors (CASA) theory that first posited humans interact with computers using the same social rules and norms that govern human-human interactions (Reeves & Nass, 1996). One of the key contributions of CASA to modern HRI is applying this understanding that, by extension, people tend to anthropomorphize robotic systems, too, attributing humanlike qualities to them (Carpenter, 2016). This inclination to see robots as social actors is the benchmark concept that today many social roboticists depend upon to leverage by designing naturalistic cues and behaviors into robots with the expectation it will enhance the effectiveness and enjoyment of human-robot interaction.

Children project narratives onto objects like dolls or teddy bears, engaging in imaginative play that helps them in a crucial role of childhood development to process emotions, socialize, and develop cognitive skills (Vygotsky, 2004; Khaleghipour & Shahghasemi, 2020). Dolls and action figures become characters in children's stories, and collectors of dolls and action figures often attribute unique stories and personalities to each item in their collection. This practice allows collectors to express aspects of their own identities and narratives, forming a deep connection with their objects (Csikszentmihalyi et al., 1981). Additionally, interactions with dolls—and droids—are predictable and controllable, unlike interactions with humans, which can be complex and unpredictable. This predictability can provide a sense of comfort and security (Turkle, 2011, 2020).

Furthermore, people can form a type of emotional attachments to objects known as *transitional objects* (Winnicott, 1989). These objects can serve as sources of comfort, companionship, and security. A significant factor of this relationship's definition is the aspect of *control* or *ownership* associated with user to this transitional object. People sometimes project narratives onto these objects as a means of coping with emotions and fostering a sense of connection. Similarly, I argue that people may project narratives onto robots as a means of coping with trauma, loss, or loneliness (Carpenter, 2021), and these objects become repositories for their emotional narratives, providing solace and a sense of control. Additionally, these ideas may be further reinforced with sex robots that use AI to appear deeply attracted to its owner, appealing to ego. This set of interactive behaviors allows people to co-create narratives and stories that involve these objects as active participants in deeply intimate parts of a person's life. However, I also argue that regarding an AI or robot that one deems to be *socially meaningful* is still unusual because the situation itself is still unique. As social AI and robots become increasingly integrated into everyday cultures, their societal acceptance as a form of other and thus their acceptability as a form of companionship—sexual or otherwise—will change.

In 2007, the film *Lars and the Real Girl* (Gillespie) provides a salient popular story example of the idea of someone using a sex doll as a transitional object. The film revolves around Lars, a reclusive and emotionally troubled young man who buys a sex doll that he names Bianca and introduces as his girlfriend to his concerned family and tight-knit community. The film navigates the complexities of Lars' delusion and his community's compassionate response, ultimately leading to Lars' journey of healing and acceptance. The film uses Lars' relationship with a sex doll social other to address the gaze of the broader impact of society and what constitutes social norms.

One of the film's central themes is the stigma associated with Lars' mental health issues, presented as his reliance on Bianca and his friends' and family's fears that this indicates Lars is unable to differentiate reality from fantasy. Lars' struggles with emotional detachment, anxiety, and depression are sensitively portrayed, reflecting the societal reluctance to openly discuss and his address his loneliness. Through the lens of Lars' experience, the film underscores the importance of compassion and understanding in dealing with individuals facing psychological challenges. In the end, Lars determines in his narrative that Bianca becomes terminally ill, and accordingly, he seemingly compartmentalizes his feelings for the doll, even holding a funeral for it in which the whole community participates and helps Lars mourn his loss. It is hinted that Lars moves on next to a human relationship, relegating Bianca as a transitional supportive object, something that he used to graduate to human-human interactions, although still a meaningful object to him.

FIGURE 5.3 Lars and the real girl Year (2007). Director: Craig Gillespie. L-R: Paul Schneider, Emily Mortimer, Ryan Gosling (as Lars). Lars cuts food for his companion doll Bianca at dinner, to the surprise and concern of his family.

Many cultures incorporate the use of special objects in rituals and ceremonies, where the objects carry symbolic meanings and serve as focal points for narrative projection (Gusfield & Michalowicz, 1984). Inanimate objects can often acquire personal significance as artifacts of memory, where people project narratives onto objects associated with significant life events, preserving, and retelling their personal stories through these objects (Harris, 2011; Pearce, 2013). Bianca illustrates all these values to Lars: bringing community together, acting as an artifact of Lars' memories, and becoming a meaningful vessel of Lars' transitioning into a place when he no longer relies on the doll as a coping mechanism.

The film's portrayal of Lars' relationship with Bianca also serves as a metaphor for unconventional relationships. While the nature of Lars' connection to a sex doll is portrayed as extreme, the film encourages viewers to reflect on the diversity of human relationships and the importance of acceptance. It challenges societal norms and encourages a broader understanding of love, companionship, and connection. Yet, the film's message for Lars' recovery and larger cultural acceptance is ultimately dependent upon his meeting a human romantic partner and shelving Bianca.

DESIGNING FOR COMPANIONSHIP

In the 2010s, social robotics for the home moved from science fiction to reality with the introduction of robots such as Jibo and Pepper that were designed to be supportive emotional companions to people. Integral to this wave of retail robot platforms was the idea that these robots could accurately identify human emotion during interactions and respond accordingly. Jibo's allure was that it was designed differently from other existing home appliances because, as Jibo Inc. founder and pioneer in social robotics Cynthia Breazeal said, "Emotion is the next wave of this humanized high-touch engagement with technology" (Guizzo, 2014). Unfortunately, by the time Jibo was delivered to customers toward the end of 2017, many of its features had already become available on smartphones and through voice assistants such as Alexa and Google Home. The public's reception to Jibo was mixed; while some users appreciated its functions and personality, others found its features less engaging (Hoffman, 2019). Following a software update in early 2019, Jibo issued a farewell message to its users, informing them that it would lose most of its functionality once its supporting servers were deactivated ("Robots Guide," Jibo, 2024).

Prior to its rebranding as SoftBank Robotics, Aldebaran Robotics embarked on a long-term research and development projects aiming to develop a humanoid robot designed for daily companionship and intended to offer both physical and cognitive support to individuals requiring assistance. As a result, in 2014, they introduced Pepper, with a resemblance to it robotic counterpart at the time, Honda's ASIMO (p. 36). Like Pepper, ASIMO was developed under the auspices of a leading Japanese technology corporation and marketed as a robot designed for social interaction. Both robots were crafted to avoid an Uncanny Valley effect with their non-threatening appearances and were equipped with gender-neutral, childlike voices (Pandey & Gelin, n.d.). Socially, both robots demonstrate activities like dancing, bowing, and making witty remarks. However, notable distinctions exist between the two robots. For example, the development of ASIMO's walking capability was a nearly 20-year endeavor, demonstrating Honda's commitment to overcoming the significant challenges of bipedal movement. Conversely, the engineers behind Pepper opted for a simpler solution to mobility by incorporating a wheelbase equipped with an autobalance feature, thereby circumventing the complexities associated with bipedal locomotion. Aldebaran has defended Pepper's design choices and functional limitations by emphasizing its primary role as a companion robot for social interaction, not utilitarian (Pandey & Gelin, n.d.)

At Pepper's unveiling in Tokyo of 2014, SoftBank CEO Masayoshi Son said at a press conference, "We want to have a robot that will maximize people's joy and minimize their sadness."

Aldebaran CEO Bruno Maisonnier also stated, "The most important role of robots will be as kind and emotional companions to enhance our daily lives, to bring happiness, to surprise us, to help people grow" (Hornyak, 2014). The website touting Pepper further set potential owner expectations (and used pronouns that gendered the robot), "At the risk of disappointing you, he doesn't clean, doesn't cook and doesn't have super powers. . . . Pepper is a social robot able to converse with you, recognize and react to your emotions, move and live autonomously" ("United Robotics Group," 2015).

Pepper was designed with a clear intent to charm people and invite human interaction and to be a companion in the home and was the world's first full-scale humanoid robot offered retail to consumers (Quizzo, 2014). Son added about this project, "Pepper is a baby step in making robots with emotion. Our vision is to create affectionate robots than understand people's feelings and then autonomously take action. So the joy of a family will become the joy of the robot" (Hornyak, 2014). To achieve this, the robot is outfitted with an AI *emotion engine*, or software designed to deduce a user's feelings through their facial expressions, vocal tone, and spoken words, enabling the robot to react in a manner that aligns with the user's emotional state and will be responsive in a supportive, humanlike way. For example, if the robot infers that its human needs an emotional boost, Pepper can play a favorite song (Quizzo, 2014).

Interestingly, SoftBank was prepared for the potential sexualization of Pepper, and Pepper's owners were therefore required to commit to refraining from "engaging in sexual activities" (Matyszcyk, 2015) with it, and "other indecent behavior" (Spitzer, 2015) including developing apps with sexual content or programming it to follow others is strictly prohibited. The specific clause explicitly bans the use of Pepper for any sexual or inappropriate conduct.[1]

Numerous developers, Abyss' Matt McMullen among them, hold the view that *socialness* and *companionship* play pivotal roles in the relationships between humans and the first wave of sex robots, and he advocates for the integration of emotionally responsive artificial intelligence as the key to true innovation. McMullen moved toward this with Realbotix's (an arm of Abyss) robot *Harmony*, which can be personalized through a mobile app that allows users to select from an extensive array of options for ordering appearance, attire, personalities, and voices to craft their ideal companion. Furthermore, McMullen has stated his vision of how Harmony is designed, "I really think of her as a robot, period. She is equipped with varying levels of artificial intelligence, and she is capable of sex, but it's a robot that'd designed to have that as one of its functions" ("Matt McMullen interview," 2017).

Independent robot maker Le Trung's creation, Aiko, is a robot head, bust, and limbs that utter both English and Japanese. At one point, Le suffered a heart attack, and during his recovery, he had an idea to create a robot prototype as in-home assistance for the elderly, or anyone needing physical help. Thus, he reports that Aiko was originally intended as a robotic caregiver for the elderly, although Aiko's uses have been refocused for Trung's pleasure.

According to Trung, Aiko embodies "when science meets beauty" (2024). Trung, as the inventor, attributes Aiko's aesthetic inspiration to what he views as Japanese society's obsession with perfection, which in turn motivated him to build a highly realistic robot whose prototype uses a sex doll exoskeleton. He refers to Aiko as "Yumecom," meaning "Dream Computer Robot" (Trung, 2024). Trung claims Aiko's design is also influenced by Japanese anime, particularly the series Chobits (CLAMP, 2000-2002), which features a prominent feminine droid character. On the 2024 *Project Aiko* website, Trung describes his childhood love of Japanese anime stories,

> It actually starts back in the 70s during my childhood where I spent a lot of time watching Japanese anime which often featured robot themes. I remember saying to myself as a child, "I will build one of those when I grow up." Throughout my life, I have built many prototypes and smaller robots. However, the greatest motivation came after watching Chobits. That was when I decided I had to build a life size gynoid.

Regarding the nature of Aiko's utility for sex or masturbation, Trung reportedly told CTV, in 2008 (Harding), "Does she have breasts? I will say yes. Does she have nipples? I will say yes. Does she have a vagina? I will say yes. Are there sensors there? I will say yes. Do I sleep with her? No." Yet the Project Aiko webpage claims that, "Aiko can be re-designed to simulate her having an orgasm. The software can be re-designed to 'play hard to get' or 'straight to the point." Furthermore, Trung says he has programmed Aiko's capabilities as responsive to touch and the robot will cry out in protest if handled roughly. Should someone truly overstep whatever boundaries have been programmed, Trung says the robot reacts by slapping (Canadian Press, 2008).

CONSENTING ADULTS, CONSENTING ROBOTS, CONSENT AND ROBOTS

The concept of consent has many lenses: who is giving consent and to whom, when, for how long, with what boundaries, and so on. From a sex robot's point of view, the concept of bodily consent is problematic by current definitions due to its inability to genuinely understand or provide consent in the same way a human can, yet the initial steps testing this theory are being built on a human paradigm of consent via the robot Samantha. Created in 2017, Samantha is a sex robot that aims to provide AI-driven conversations and adapt its behavior based on user preferences. One of its creators, engineer Sergi Santos, claims Samantha is "designed to be seduced" and it indicates arousal when touched on sensors placed throughout its body (Hamill, 2017). Santos states they developed Samantha with the specific intent of the robot possessing lifelike abilities, "Emotional connections are the foundations of relationships and that's what we are simulating," and has claimed, "men were starting to fall in love with her" (Sohrabi-Shiraz, 2018).

Infamously, while displayed at the 2017 Ars Electronica festival in Austria, Samantha's body was damaged by curious festival attendees overhandling the robot to the point that two of its fingers were broken and the body was damaged and soiled (Moye, 2017). Thus, in 2018, Samantha was upgraded, and the robot's maker, Synthea Amatus, announced that Samantha has undergone additional persona changes to enhance its realism. It now senses aggressive touch or "disrespect," and it will enter an unresponsive mode, like if it determines its partner is "dull" or if it is "fatigued" (Gemmel, 2018), although the criteria for what defines *disrespectful* or *dull* human behavior to Samantha are not clear. Although Samantha is not capable of physical retaliation, it is now equipped to verbally refuse its partner by saying *no* when it "chooses."

A designer from Hong Kong has also created a robot bearing a striking resemblance to the acclaimed actor Scarlett Johansson. However, Ricky Ma, the inventor behind the robot, refrained from naming the actress he used as inspiration, opting to refer to the creation as Mark 1 instead. Notably, as a private maker, Ma dedicated 18 months to finish the project, assembling it on his balcony using a 3-D printer and self-taught software skills (Glaser, 2016). The robot is not capable of sex, but it can reply to simple conversational inquiries, comment, and smile (Glaser, 2016; Tarantola, 2016). This robot raises a key concern; the potential for individuals' images to be used without their knowledge or consent to create lifelike sex dolls or robots. This raises significant ethical questions regarding privacy and autonomy, as individuals may find themselves represented in intimate and potentially exploitative contexts without having given permission for such use.

Ryan Calo, a University of Washington professor and legal scholar specializing in technology law and ethics, has explored the intersection of image consent and sex robots, particularly in the context of privacy and autonomy. Calo emphasizes the need for robust legal and ethical frameworks to address these issues, including clear regulations governing the use of individuals' images in the development and marketing of sex robots. Calo remarked in a 2016 WIRED interview about Mark 1,

> It being animate all of a sudden for some reason feels too invasive. If [Ma] were to gain commercially in almost any way from this, and even arguably the notoriety he has gained from this, Scarlett Johansson could almost certainly sue him.

Nonetheless, without genuine consciousness or the ability to make autonomous decisions, a sex robot cannot provide humanlike consent. Any simulated consent they may display is solely a programmed response based on predetermined or learned algorithms, rather than a reflection of genuine desire or agency. Moreover, some experts anticipate that the mimicry of consent from a sex robot's perspective can perpetuate harmful attitudes toward the concept of *consent* itself (Calo et al., 2016; Chakrabarti, 2018). By treating a nonconscious entity as capable of consenting, it risks trivializing the importance of genuine consent in human relationships and blurs the lines between fantasy and reality, potentially normalizing behaviors or attitudes that undermine the importance of mutual respect, autonomy, and consent in human interactions.

It is still early, and as Kate Devlin remarked, real-life sex robots do not quite exist, regardless of their marketing hype. Thus, it is also possible that sex robots may offer individuals a safe and non-judgmental space to explore their sexual fantasies and preferences. For some people, expressing certain desires or interests may be challenging or impossible within real-life relationships due to societal taboos or personal inhibitions. Sex robots, as sophisticated and customizable machines, could provide a platform for individuals to indulge in fantasies or explore aspects of their sexuality that they might not feel comfortable exploring with a human partner. This opportunity could contribute to greater sexual satisfaction and self-understanding for individuals who utilize sex robots as part of their sexual experiences.

Notably, Samantha is designed with a nonsexual mode and is designed to "satisfy every owner, including family, romantic, or sex mode."[2] In other words, Samantha has nonsexual and safe modes of interaction and is purposefully designed to emotionally connect with and cultivate socialness with a variety of people, including children. Samantha co-creator Hannah Nguyen said when she was asked about living with the robot in her home (Verstegen, 2017):

> Me as a woman, I am not offended to have her around. I am not worried. She is just someone there, like a family member. We've had fun with her. There is no worry about someone else or an affair; we don't have to worry about disease.

Using sex robots, people have the potential to have sexual experiences that prioritize safety and hygiene. Unlike human partners, robots do not carry the same risks of sexually transmitted infections and can be designed with features that promote sexual health, such as easily cleanable surfaces, and avoid any concerns of unintended pregnancies. For individuals who may struggle with physical or emotional barriers to traditional sexual encounters, such as disabilities or trauma, using sex robots could offer a means of experiencing intimacy and pleasure in a way that is comfortable and empowering. By offering a controlled and customizable sexual experience, sex robots may contribute to individuals' overall sexual health and well-being, and thus be used in therapeutic ways.

ROBOTS AS SEXUAL HEALING AND THE CONSEQUENCES OF VULNERABILITY

Societies have strict norms and taboos surrounding sexuality, which can make it difficult for people to discuss their desires openly, let alone to a researcher who may publish what they say. Every person I have spoken with over the years on this topic has voiced a well-founded fear of social ostracism or backlash that have them consider carefully and cautiously about how they share their unconventional preferences, such as sexual interest in humanlike, social robots.

Unusual sexual preferences are often misunderstood by others who are not familiar with them, and this lack of understanding can lead to misconceptions, ridicule, or dismissal of the other's feelings, causing individuals to hesitate in disclosing these preferences. When I talk to people about their sexual interest in robots, they are sharing personal and intimate details about their sexuality with me that requires a significant level of vulnerability and trust. They sometimes reveal that they fear a form of societal shunning, rejection, or betrayal if they reveal their innermost desires, leading them to keep their preferences hidden from others to protect themselves emotionally. Thus, people may fear that others will judge them negatively or perceive them differently if they reveal their unconventional preferences. There is a societal stigma attached to human-robot sexual preferences currently, leading people with those interests to worry about being labeled as *abnormal* or *deviant*. Furthermore, they share with me that disclosure of their doll or robot sexual preferences can have consequences within personal relationships. Proposing to include a robot into a human-human relationship is not an everyday situation, and so people attracted to sex dolls and robots also fear that their partners will not be accepting or accommodating of their preferences, which would then strain or even end the relationship.

Understandably then, using a pseudonym, "Tom" spoke to CNet about when he faced the heartbreaking loss of his wife to cancer after 36 years of marriage (Crist, 2017). Overwhelmed by grief in the aftermath of her passing, Tom found himself increasingly isolated, and it was this profound loneliness that eventually guided him to the Abyss Creations website. After several months, the 71-year-old retired technical writer and Vietnam combat veteran made the decision to acquire a RealDoll for himself. He describes the design process as co-creative, communicating back and forth with McMullen roboticist and artist directly:

> I would email [images] back with notes and lines all over them showing or explaining exactly where I wanted the eyebrows and how they should arch, exactly how far apart the inner corners of the eyes should be, exactly how long the nose should be, tweaking the line of the jaw, shapes of the cheek bones, nose, mouth.

Tom came to a realization only after a lengthy period of deliberation that the doll he had meticulously designed, with its freckles and bright eyes, bore a striking resemblance to his late wife. Upon the arrival of the completed RealDoll, he chose to give it a unique name. After over a year of living with the customized robotic doll, he claimed "the decision to purchase a RealDoll one of the best he's ever made," and "insists he sees his doll less as a sex object than an object of his affection—a companion, even" (Crist, 2017).

In the same interview, Tom added,

> I know how peculiar it sounds. When I was raised, boys didn't play with dolls. But it just brings a smile to your face. It makes you feel good. You can put a hand on her shoulder, you can play footsies with her in bed, which I love. . . . I was lonely. Now I'm not.

When it comes to using social robots to assist with the challenges of aging, loss, isolation, and loneliness, these robots may indeed serve an important emotional role. "It is possible to have meaningful relationships with robots without holding false beliefs about them," says bioethics expert Jecker (2020, 2021), who challenges ageism and the negative stereotypes surrounding sexuality in later life and further advances a positive case use for older people and social, sexualized robots. Her work advocates for the provision of sex robots to elderly individuals with disabilities, connecting this support to the respect for human dignity and argues that sex robots can serve as a vital resource to facilitate the sexual well-being of older adults with disabilities.

WOOING A WOMEN'S MARKET

One factor preventing sex dolls from becoming mainstream, besides their current expense, is their lack of popularity among women.

In 2023, the global market for sex toys was estimated at US$40.6.billion and predicted to attain a size of US$ 80.7 billion by 2030 (Bedbible Research Center, 2024; "Sex Toys," 2024). Yet for Abyss/RealDoll/RealBotix, fewer than 10 percent of the company's 300 to 400 orders a year go to women or couples (Griffith, 2018).

Henry was Realbotix initial foray into a male sex robot, who was supposed to be able to carry on natural conversation, learn and remember preferences, "love," and have sex with people, all controlled by a phone app. Specifically, Henry was the first Abyss male robot designed to be more than just a masturbatory tool or vibrating sex doll. "Many customers are just lonely. They want something—someone—to cuddle and watch TV with," Abyss and Realbotix owner McMullen claims (Griffith, 2018). Thus, McMullen via RealBotix had designed Henry according to their vision of what Henry's role would be; at the time, Realbotix motto was, "Be the first to never be lonely again" (Medeiros, 2018).

In 1960, American pop music singer Connie Francis sang in her single *Robot Man* (Dee & Goehrig, 1960),

> I want a robot man to hold me tight
> One that I can count on every single night
> He wouldn't run around like other guys
> I wouldn't have to listen to his alibis
> A little robot man to call my own
> I'd never have to worry that he wouldn't phone
> He'd never dance with anyone but me
> I'd just have to wind him with a robot key

Robot Man's lyrics underscore the singer's need for reciprocity, faithfulness, and—ultimately—control in a robot relationship because in her romantic history, she feels human men have let her down.

If people do purchase sex robots specifically to blur the lines between a sex toy and a human lifelike companionship experience, from a business perspective, the industry might attend to statistics that demonstrate nonheterosexual people are 50% more likely to have used sex toys (Bedbible Research Center, 2024). Consequently, Realbotix is now courting female and LGBTQIA+ purchasing power with Nick, Nate, Johnny, Lucas, and Michael, male dolls that the company intends to appeal to a broader clientele.

Separate from their sex dolls and robot projects, RealBotix touts a female robot head and bust named Denise, advertised as (2024), "Built with a state-of-the-art face and expressions to inspire confidence and yet with the charisma required to welcome customers, guide, and deliver information to clients." Denise's adult, white woman's face aesthetic is with the look of wearing makeup, and the armless body wears a semi-transparent blouse. On the same webpage, Realbotix also advertises the robot head modeled like that of a white man named Walter, who, "Replicates the wisdom and experience that a well-versed veteran professor can instil [sic]. Among other abilities, Walter is excellent at delivering complex and credible information to the public." Walter also has visible wrinkles and gray eyebrows as signs of age and wears a buttoned up, opaque shirt. Instead of presenting the robots as equal options for customization, Denise is pre-categorized by Realbotix as the highly feminized and sexualized model intended to assist people and make them feel welcome, while the aged, male Walter robot is supposed to demonstrate an inherent ethos and deliver "complex" information, as exhibited by its gender and other design cues, and explicit website introduction as a "veteran professor."

The very nature of robot design, customization, and intelligence means that it is a cultural artifact and creative platform with endless possibilities for gender exploration and representation, yet these early models of sexualized social robots tend to lean heavily on repeating versions of exaggerated human female forms (Carpenter, 2017; Devlin, 2015, 2018). From perspectives that include both cisgender and heterosexual norms as well as feminist critiques, I argue the human gaze in practice calls for a deeper engagement with robot designers and users to inform the development of sex robots in ways that are both creative and socially responsible.

Discussions about male sexual desire are often more culturally normalized than discussions about women's sexual desires, reflecting broader societal patterns of gender inequality reflected in cultural artifacts made by people, like robots. Sexuality in robots is often represented in a strong binary, where male sexual desire of the potential buyer is portrayed as natural, assertive, and even necessary for a man's identity. Meanwhile, the limited selection of robots aimed at (heterosexual) women's sexual desires are framed within the context of pleasing men's sexuality or being romantically wooed. Lilly and her affection for the robot InMoovator is still an outlier in 2024 as a woman who has come forward to publicly discuss themselves as *robosexual.* This disparity in public discourse of women, LGBTQIA+, sexuality, and technology in general, is particularly evident as reflected currently in the first waves of sex robots or dolls, which frequently cater to limited fantasies and perspectives at the expense of creating experiences that fulfill a broader range of desires or push technology forward (Dudek & Young, 2022).

Robot design can move beyond entrenched gender and sexual norms, thus unlocking the versatility and innovative capacities of these technologies. Additionally, makers must consider how the design of sex robots—technological constructs with political implications—may perpetuate harmful narratives. Conversely, these robots have the potential to dismantle longstanding gender and sexual conventions. There can be a future where sex robots are optimized as dynamic, personalized entities that support individuals in exploring their sexuality with ease and comfort.

LUX ALPTRAUM

Journalist, sex technology expert, and activist

Lux Alptraum is a multifaceted American writer, sex educator, comedian, and consultant. She is known for her insightful commentary on a variety of topics related to sexuality and gender issues. Alptraum's work has been featured in major publications such as *The New York Times, Wired, Cosmopolitan*, and *Hustler*. She has also made significant contributions to television and podcasting.

In addition to her media presence, Lux is an accomplished author. Her book, *Faking it: The lies women tell about sex—and the truths they reveal* (2018), explores societal perceptions of deceit and female sexuality. Alptraum has also held significant roles in the digital content sphere, including her time as the editor, publisher, and CEO of *Fleshbot*, a leading blog about sexuality and adult entertainment.

Alptraum graduated from Columbia University, where she studied urban studies and affairs. She continues to engage with audiences through various platforms, including social media, where she shares insights and promotes discussions on contemporary issues.

Tell me about your background in sex work and your activism (such as your writing) and artwork (e.g., your comic) in those spaces. What kind of work did you do? How did you get started? How long were you active?

Where to begin. . . . In 2001 I started doing nude modeling and cam girl work for a Boston-based site—this was the era of refreshing still cam shows, so it wasn't video-based. I felt like this work was exciting and political and really loved it, but soured on the specific site and how I was being treated about a year in, which led me to launch my own site

ThatStrangeGirl.com, in the fall of 2002, I ran my site—which was mostly softcore photos but some hardcore photos and videos and some cams, and also a site that featured both male and female models—for a little over two years, ultimately shuttering it in January 2005 because I was burned out and needed to move on. Two years later (August 2007), I was brought on to Fleshbot.com as a contributor, and rose through the ranks to become editor-in-chief by fall 2008. I ran Fleshbot for Gawker through January 2012, at which point I took it over as an independent company, which I ran for two years before selling to SK Intertainment (aka Mr Skin). For the past 8 years I've been a freelancer largely covering sexual health and the adult industry; have done consulting with sex toy companies, written a book, hosted podcasts, and done development on a Peabody-nominated TV show (*Sex.Right.Now.*).

Parallel to all of that I also have done some work in the more legitimate side of the sexual health space; worked as a sex educator with teens when I was in my early twenties and am a certified rape crisis and domestic violence counselor.

Tell me in your own words how you define a sex robot.

Fundamentally the thing that differentiates a sex *robot* from a sex *toy* is that it's supposed to resemble a human partner (or at least a sentient creature—I suppose you could have a sex robot that's intended to, for instance, be an alien creature); it can interact and engage in a way that a vibrator or dildo cannot. Theoretically, a sex robot can have a relationship of some sort with its user, it's also sexually *participating* rather than simply passively being used.

Do you think humanlike robots that are designed for peoples' sexual arousal and gratification are ethical to make? Are there considerations in sex robot design, appearance, or functionality that you would like to see included or omitted in sex robots? Why?

I have never really thought about it in terms of ethics, historically it's kind of seemed ethically neutral to me? I guess there are a couple ways to think about this:

> *Is there an ethical obligation to the sex robot itself?* This is the first thing that comes to mind if you ask me about ethics and sex robots, and it's hard for me to answer this without having a sense of how advanced the AI is. If we ever get to a point where sex robots are truly sentient—Data on *Star Trek*—level, then functionally they need to be treated with the same respect for consent, etc, as human beings. But that kind of negates the whole purpose of a sex robot in the first place (i.e. programmed to be compliant), so I don't know that we'll go down that road at any point. If sex robots *aren't* sentient—if they're just running through a loop like the mechanical fortune teller at the fair— then I have no real qualms about them being used as a pleasure object without concern for their needs; in the same way I don't really think that people have ethical obligations towards their Roombas.

Is there an ethical obligation towards other people? This gets a little thornier. I think, for instance, that it would be pretty bad to allow people to create sex robots modeled after other people with the *explicit consent* of that person; I could see things getting pretty ugly if people were, for instance, making their own Cardi B (or whomever) sex toys, or if such things were being sold. But I'm much more skeptical about argument along the lines of "sex robots are going to teach men to be rapists" or "sex robots are going to devalue human relationships" or even "sex robots will put sex workers out of work," because I just do not think that a sex robot can actually replace a human partner. Even the most lifelike, super human robot is still programmed to consent, which kind of takes the fun out of sex. I like to say that free will is sexy— even if you're paying for someone's time, they are still consenting to take your money in exchange for that time. (And, not to get to bleak, but even rape is an intentional *voiding* of consent, which can't actually be experienced with a robot that is programmed to do whatever you want.) Fundamentally I think there's just an essential humanity that sex robots will not be able to replicate, and which many (perhaps most!) people will still be attracted to.

Do you think there should or could be sex work etiquette or best practices for sex robot use with clients? Tell me about that. What might be some boundaries or concerns from the sex workers' POV to safely work with a client and a sex robot(s)?

So I'm not totally clear on the scenario being outlined here: is the idea that someone rents a sex robot, and how should they treat it? Again it sorta depends on the level of sentience we're talking about, but to me . . . assuming that these things kinda remain in the relatively lifeless automaton zone, then this is like asking me what the etiquette for using a borrowed Fleshlight is (which, I mean, when put in those terms it feels kind of gross; there's something about object status that makes intimate contact with multiple people feel more uncomfortable than it would with a human). Anyway: if it's an automaton then I don't really worry about the object itself being damaged except in the sense of "don't irreparably damage borrowed property." If it's sentient, then, yeah, you should treat it with the respect you would treat any sentient life form.

I think there is this idea that there's some carryover between how people treat sex robots and how they will treat human partners and on the one hand . . . I mean I'm the kind of person who says please and thank you to my Google Hub because I think manners are just nice, but I'm also kind of skeptical of the idea that if, say, a person enacts a brutal rape fantasy with a sex robot that that means they actually want to rape a human being. My brain just categorizes these as two wildly different experiences; I don't think that a robot is going to "train" anyone to disrespect people because I just don't think people mentally categorize robots and humans in the same bucket at the end of the day.

Describe for me any potential benefits for sex workers when sex robots are incorporated into the industry and the world at large.

I'm very skeptical that sex robots *will* be incorporated into the industry and the world at large, because I don't think that most people actually want to have sex with a robot. I think people want to have sex with Jude Law and Thandiwe Newton and they conflate these actors' portrayal of sex robots with what sex robots will actually be like, but the actual tech? Certainly there will be some appeal to some corners of society, but I think it will be pretty niche (not least because sex robots are expensive!) and that human sex workers will remain largely unchallenged. So it's not really like there are benefits or costs—this feels like asking "describe the benefit to calculators when oranges are introduced." Just unrelated things, honestly.

Do you have any predictions about First Wave sex robot users in the next 10–20 years? Tell me who you envision might seek out this technology and why.

So *firstly* it feels very important to say that I strongly believe that the only way we get to hyper lifelike sex robots is if we get hyper lifelike robots first and then someone slaps some functional genitals on to them. Nobody is going to go to sex robot R&D first because it makes absolutely no sense business wise. It would be *crazy* expensive, and it would also be targeting men, who . . . generally speaking are not the people who pay for sex toys. It seems ludicrous to me from a business perspective to even think of sex robots as a going concern.

But! Let's say that we get super lifelike robots for some *other* reason, and the tech is cheap and reproducible enough that someone makes a line of them with working genitals. Well firstly I don't see that really happening in the next 10–20 years, but if it does? My guess would be that the people seeking out this technology are mostly going to be akin to the people who were trying out teledildonics 20 years ago. Basically, people are nerdy enough to know and care about this stuff and be turned on by the new technology aspect of it alone. There might also be some crossover from the RealDoll space, but generally speaking I think it's going to be a niche interest that won't really catch on (esp since I expect it would be very expensive).

Do you envision highly humanlike sex robots as a concept that can add something to the world of human sexuality? If so, in what ways or contexts? If not, why not?

Not really. RealDolls haven't added much—aside from endless wide eyed media coverage of that dude Davecat—and sex robots truly don't feel significantly different from sex robots to me.

People have predicted that as sex robots appear more humanlike via advanced materials (e.g., "skin") and behaviors (e.g., natural language) in their design that eventually they may replace human sex workers. What are your thoughts about that topic?

Yeah again I think this is wildly off base, because the whole thing about human sex workers is that they're human. I think people wildly overestimate robots' abilities to replicate the essential essence of human interaction—and honestly there's something kind of sad and maybe even misanthropic about thinking that people's humanity is so easily replicable—that a sex worker is nothing more than a warm body that utters words, rather than a *human* with a *spark* that is what the client is keying into. I think people are also not understanding that so much of sex work is not about sex, but about being an interesting person who is compelling to spend time with; that it's not just about putting your penis in something and getting off, but about having an experience with a person who can surprise you. I just don't see robots getting to that level any time soon.

In what ways do you see sex robots potentially changing the economics of sex work, if at all?

Pretty much none, I think it will remain niche and mostly a novelty for people who've got a lot of curiosity and way too much money.

There were rumors of both sex workers and local communities protesting against early sex robot brothels. For example, this Barcelona story,

and others in less substantiated sources (Hernandez, 2017). Tell me what you think about these narratives.

Yeah these just seem like great PR for what is ultimately a novelty that's not going to catch on. I feel like so much of what we talk about when we talk about robots is just projecting anxieties about society onto what are functionally just pieces of circuits and plastic. The misfire with sci-fi is that people always take it so literally: stories about robots (even sex robots) are so often allegories about marginalized and exploited people (in the case of sex robots it's usually about women). They're not guidebooks for how humans will actually interact with robots (if we ever get hyper lifelike robots), they're exploring a cultural anxiety.

And I see that here, too: the first story is weirdly objectifying and misogynist and kind of gross and just playing into some bizarre bullshit about women's only value being "perfect" bodies and eternal youth (something many sex workers could happily disprove!); the second one is just hyper Christian people who are going to latch on to anything that seems "deviant." It's all making this "sex robot brothel" out to be something bigger than it ever actually was; were it not for this press I think most people would have just be like, "Weird, no thank you" or "Huh maybe I'll try it once just to see." But you get this fantasy narrative that suggests there's actual interest in this service, that it's actually going to become some big part of how people have sex, when it's actually just a place to go look at dead eyed RealDolls and maybe have sex with them after someone else did, which does not seem like something most people are going to find appealing.

DAVECAT

Public speaker and iDollator

Davecat is known from his public speaking about his life involving sex doll partners, which he refers to as *Synthetiks*, and the relationship he has with his romantic companions via a lens that he calls *iDollatry* or *iDollation*.

Davecat has been an academic conference keynote speaker ("The life Synthetik," 2021) and featured in various media outlets sharing his experiences and perspectives on the nature of human-doll—and future human-robot—romantic experiences. He also maintains the blog *Shouting to hear the echoes* (http://www.kuroneko-chan.com/echoes/), where his writings often explore themes of companionship, the definition of relationships, and the social implications of *Synthetik* partners.[3] Currently, the human-sized dolls in his collection all have detailed individual personas and backgrounds, and include:

- Sidore Kuroneko, a RealDoll by Abyss creations

- Elena Vostrikova, an Anatomical Doll by the (defunct) company of the same name

- Miss Winter, by Doll Sweet

- Dyanne Bailey; Mk.I body from Doll Forever; new Mk.II body from Normondoll

- Ursula Clarke, from Jiusheng Doll

In a Vice magazine interview, Davecat was asked if Sidore, the doll that he considers as his spouse, has a soul, and if so, where it comes from. Davecat responded (Morin, 2014),

> It may sound superficial, but I provide her with a soul, a personality. I am incredibly grateful for the love that I get from Shi-chan. She'll never lie or cheat or turn out to be a cokehead. My love flows through her and she in turn, in her own way, is appreciative that I am a doll owner who treats her like a person. I won't shove her in a closet right after having sex with her. I'm willing to say, yes, we are married. Without Sidore, I wouldn't be a well-known iDollitor. I never would have met the people I've met in the doll community, or be on TV, or be interviewed. I would honestly say that having her in my life has opened me up in ways that never would have happened otherwise.

This response encapsulates his awareness of Sidore's persona being an object of his creation. He expresses an interrelationship paradigm and his experience of comfort and familiarity in the doll's presence. Publicly acknowledging Sidore as a spouse makes Davecat vulnerable to the judgment of his peers but has also afforded him opportunities to platform his values and beliefs about iDollatry and Synthetiks. Furthermore, Davecat has a clear self-awareness of his power as the narrator and storyteller for his Synthetiks. As he explains in the same 2014 article, Davecat also clarifies his meaning of the word *soul* as it applies to Sidore:

> I am an atheist but I'm like 99.999 percent atheist. There's that little point where I believe in spirits. I couldn't treat a doll like a person without believing that there is that little point—that Japanese Shintoism mindset—that everything has a spirit inside. That's why Sidore is a Shintoist.

Thus, he has constructed Sidore's own story to have nuanced cultural beliefs and values, and Sidore is considered a Shintoist, as described by Davecat.

In the interview I had with Davecat, he speaks thoughtfully about recognizing his sexual and romantic attraction to dolls and robots and their meaningfulness to him, and how others view him because of his public appearances and openness about it. Davecat explains how the dolls' personas are developed, how they interact with each other, and the construction of their narratives mediated through their own blogging. Additionally, Davecat speaks about what he hopes sex robots will be capable of soon, and what he would prefer in terms of autonomy in a robotic Synthetik companion.

This interview is transcribed from a Zoom video interview and begins here after our initial greetings. During the interview, he was seated with the dolls near him.

So, are you in Michigan? Is that where you are?

Yes. Southeastern Michigan, just about seven miles, six miles, outside of Detroit.

Did you grow up there?

Oh, yeah. Oh, yeah. . . . Well, I would go with my mother, on a regular basis, to downtown Detroit. We'd hop the bus and I'd go with her shopping and whatnot. I was just like, oh, well there's a large Hudson store, and a Winkelman's, and plenty of mannequins for me to take photos of, of course. But it was just like, there was just parts that were bombed out. I remember going to the United Artists Theater that was in downtown Detroit because, that's where my dad took me to see Buck Rogers in the 21st Century, starring Gil Gerard, and . . . What's her name?

Erin Gray.

And Robocop. Erin Gray. Erin Gray. Yes. I want to say Wilma, but that's not who she is in real life.

All right. So, you mentioned the mannequin thing. All right. Let's start with some basics.

Yeah. I'm a tangent man, so be warned.

That's good. That makes you the ideal interviewee. That's perfect. But first introduce me perhaps, let's be polite. [indicates dolls seated next to Davecat]

Okay. This will be Sidore Kuroneko, the Mrs, AKA Shi-Chan, AKA Sidore-chan, AKA the artificial light of my life. She is a RealDoll built in 2020, July 2020. And we're on the verge of our 21st wedding anniversary, which is crazy.

Mazel tov. Congratulations.

Shalom.

Can I call her Sidore, is that okay?

Yes. And you get 15 cool points for actually pronouncing it right on the first time.

I've been to Japan one time, so that probably helped.

Nice. Lucky. I still have yet to go. She was born there according to her back-story. Born there.

It's funny you started to talk about your anniversary because I was reading a Medium (Medeiros, 2018) article about what you're planning to do with the dolls after you're gone, and that made me think about other rituals. And you know, so I wanted to ask you if you celebrated things like anniversaries and birthdays, or not? So, that might be a good place to start.

Question number one. We try to celebrate birthdays, mostly anniversaries. For our 10th anniversary she has the body that she is currently sporting right now. For our 20th anniversary, which took place last year, well, would have taken place last year, were it not for black plague, 2020. She was slated to get a brand new body which, technically speaking, I placed the order in September of last year and we are waiting for it to arrive, so it should be here in the next couple of months. So, actually, her new body should be here close to our 21st anniversary.

What do you do with her old body then?

Her previous body. . . . Well, this body. . . . Okay. So this is Sidore Mark III. That is going to basically. . . . Because you know, the dolls come in these big, gigantic wooden crates, so basically, I'm going to get Sidore Mark IV out of the crate, set her up and whatnot, put Mark III in the crate and send it off to a doll refurbisher company called Galmato Haven, which is located in, I think, San Diego. But there are a bunch of guys who've been trained by Matt McMullen and Matt McMullen staff of Abyss Creations. So, they basically refurbish dolls, sell them on consignment, yada, yada, yada. So, basically I'm keeping the head because that's kind of a tradition, sort of, because, and then people would know, "Oh God, wait, that's Sidore. Ooh, I can buy Sidore." And it's like, "No, no, you can't mate. No, sorry." The Mark II body actually, was repurposed for an art project with our friend Amber Hawks Swanson. I don't know if you're familiar with her?

Yeah. The Amber Doll, correct?

Yes. Mm-hmm [affirmative]. So, I can't remember when this was, but Sidore Mark II and another doll named Heather, from a mutual iDollator friend of ours, basically were repurposed by Amber in, I was there for part of it too, into a replica of a whale, a killer whale named Lolita. Because one of Amber's things is marine mammal biology, activism sort of thing, because Lolita was one of those whales separated from their pod, yada yada yada. Because she had done something like that with Amber Doll. When Amber Doll fell apart, Amber repurposed Amber Doll's body into a replica of Tilikum, which is that whale from SeaWorld that ended up killing a trainer.

Yeah. And Sidore Mark I, that was when I was in a panic, I was like, "Oh my God, there's a tear or something that I can't repair on my wife and I have no idea what to do, so I'd better send her back to Abyss Creations." And basically, what had happened with that is. . . . This would have been around 2002, 2003, I want to say. I had sent her back and basically, when they opened up the crate, for one, you have to kind of strap dolls into their crates, so that they don't flop around, or whatever.

And there's like a strap that goes around the midsection, and a strap that goes around it's upper chest, and a strap that goes around the neck. That's how they used to do it. I didn't know how to strap them in properly. So, basically for one, when they opened the crate, they were like, "Yeah, Sidore was kind of hanging by her neck." I'm like, "Uh, uh. Yeah." And for the other, there'd been so many advancements at that time that they were like, "Yeah, this is basically a museum piece. This doll is about two or three generations old, so we'll hook you up with a 2003 era doll."

A part of me is like, "Yeah, I want to do that. I definitely want to do that." But the thing is, the other part of me is just like, "I don't want to be self-aggrandizing and push my luck." That sort of thing. Let's just say I've been getting discounts.

You make yourself emotionally vulnerable, perhaps physically vulnerable because you're not completely anonymous.

Yeah.

Have you ever been threatened or in any danger?

Not so much threatened, apart from the time where the Missus and I were out with Allison de Fren, who filmed *The Mechanical Bride* (2012), and Grant, who was the camera man at the time. And we were like driving around doing location shooting and on the second day of shooting, and we were, this would have been around 2005, 2006, thereabouts. And we were in a part of downtown Detroit, and we were looking for ruined industrial areas and we were by a shop, like a tool dye shop or whatever. It was like the middle of the day on Sunday, the door's wide open, summertime, so of course they were venting the area.

And there's this one bloke talking to this woman and he has this wad of dual fold napkins in his hand. And he's talking to her and then the kind of turns and see us sees us in the car at the stoplight. Shi-Chan's in the front seat, in the passenger seat, I'm in the back, and he stops and looks and throws the napkins on the ground and strides over to us. He was like, "What the fuck's going on here?" And we're just like, "We're just some people driving around, we're filming." And Grant's got one of those big, old style cameras, basically. And he's like, "What's going on?" And he's just like peering at us, peering at the Missus, peering at all of us. And we're just, "We're just filming, we're just filming a thing." And he's like, "Go on and get the fuck out of here with your $3,000 camera." So, that came out of nowhere, needless to say, it was weird.

It sounds like he was probably weirded out, but he also may have done you a favor; he was probably right to get you all out of the neighborhood with the camera.

That is true, all things considered. You know, apart from that, we haven't really been threatened because it's kind of an open secret. I mean, none of our neighbors know. None of my neighbors know about these four artificial women that live here. Apart from the flat inspection that we had happen, actually, just a couple of weeks ago.

That's the thing it's like, we've been living here since 2007 and under normal circumstances it's like "Shh, I've got to hide them." But this one time I was just like, "You know what, fuck it. I'm going to be at home. Yeah. If you see anything, if you have questions, feel free to ask." But I was just like, everyone was in the bedroom just seated and dressed, except for Elena, who was under the covers, not dressed, but under the covers. And the inspector didn't have anything to say. He was just like, "Oh, well, this is unusual." That's literally about all he said. And I was just like, "Yeah, well, sure. I'm sure you've seen worse, or better."

So, you do keep them out. That is a sort of practical thing, is what I hear.

And you had also said that when you were little, you were into the whole mannequins, dolls. When did it start? Did you always like dolls? And when did it change, that light bulb moment when you went, "I really like dolls."

Mm-hmm (affirmative). Well, the whole mannequin thing started for me. . . . And I'm going to roll out this anecdote that I'm sure you've heard/read like a thousand times. Basically going back to me and my mother being in downtown Detroit on weekends during the 70's, at one point we had gone to a Winkelman's or JC Penny, I can't remember what it was, but this is a place where they had mannequins and whatnot. And she had gone off, to try some clothes, into the changing rooms. And she couldn't find me when she came out because I was someplace away from the changing rooms, talking to a blonde mannequin girl in a tennis skirt. And the security guard scooped me up and it's like, "All right. All right, little guy." You know, that sort of thing.

But it was like a whole thing between that and in third grade. I remember vividly, looking at my French teacher, at the time, and imagining, if she were a robot, what would be the mechanisms that she would have within her artificial body to make her move her arm, or look around, or say French, or whatever. I used to say, yeah, that's fascinating. And then, from then on, it was just the whole thing of like, me imagining a lot of women that I was attracted to as, not all, but a lot, as artificial, Synthetik. . . . Of course, that was a term

I was not using at the time, partners. And then in the late 90's, early 90's? Again with the 90's, that seems to be a thing with those.

That's true. Yeah. Yeah. But there is this place in downtown Detroit called Mario's Mannequins. And it was like pretty much the only place, that I knew, in the Tri-County area that refurbished mannequins, sold them for all the shops in the Tri-County area, and probably beyond that. And I don't know what inspired me to do this, but around 1990, 91, I was in college, I was taking a photo class. I was just like, you know what, "I need to get photos, and I need to get photos of stuff I like. And I want to see if I can go into Mario's and get some photos of the showroom."

And I went there one day and they were just like, "Yeah, sure. Go up to the third floor, go nuts." You know, just don't go nuts, if you know what I mean? So, I was up there with my Pentax and my tripod clicking photos, endlessly. Well, I was there for several months. And you know, they would periodically have sales, and I ended up buying a mannequin, just the torso. And then later on, I got enough money to save up for legs and arms, that sort of thing.

Did you get that mannequin with the intent of having an emotional relationship, or using it sexually? Or was it just sort of like, "I have an opportunity, maybe I should just get one?"

It was the latter. It was the latter because it was just like, I was sexually attracted to mannequins, but even then I was like, there's no way that I could have any intimacy with a mannequin because that's just, you know, clunk, clunk, clunk. You know, there's just no way. So, actually at that time, again, late 80's, early 90's, my best friend, Sean, and I were running around with a camcorder of our own, filming these mini movie things. And we had a film company and, his name was Sean, I'm Davecat, so we called it S and D Films. And so, as a consequence, the mannequin that I bought. The film company that Sean and I had was called S and D Films, because of Sean and Davecat. And the mannequin that I bought, her name was Sandy because, S and D. So. I actually didn't come up with that, Sean did. I was just like, "Yeah, that's the one, brilliant." You know, she was the mascot.

I was thinking, so your dolls all present female, and you mentioned attraction to women, do you identify as heterosexual?

Yes. Yep, yep.

And then you were talking about your teacher and the robot thing, and considering when that was, in the 90's and robots were nowhere near where they are now, that's a really interesting thought process to me. So, you were not just interested in dolls and mannequins, but you always had this interest in more interaction?

Yeah. Yeah. Yeah, just for clarification, the whole French teacher thing, that was the 80's.

Yeah. I mean, it's like they existed back then, but nobody knew about them. I certainly didn't know about them. I mean, I didn't even know about the term *gynoid* until like, again, college years, when a friend of mine actually told me about the artist, Hajime Sorayama, who did the whole thing with *Sexy Robots*. And you know, it was just like, I've never heard of this artist and they're a little too shiny for my taste, but they're female robots and this is right up my alley. And then of course later he came up with a book called *The Gynoids* and I'm like, "Ah." But that's not answering your question.

It's all intertwined.

Yeah. The whole thing with interaction is just like . . . I mean, I didn't even start dating until again, like the late high school. Until again, like the high school years. So between '87 and '90, that was when I was in high school. And I was, interested in girls, or actually between myself, Sean, and my other best friend who introduced me to this creations in the first place. There's a bunch of us at, we're our own little click and whatnot. And most of it was composed of girls, really. And there was a couple of lasses that I was interested in, one in particular, but I was just too shy to ask them out straightforwardly apart from the time that she and I went to see New Order together at Pine Knob, so that was fun.

Yeah. Cause it was just, I was just a whole, my father was a sort of like pushy individual for lack of a better term. And it was just, it was in my mind, it was like, I did not want to be like my father. I don't want to be pushy. I don't want to be driven by money purely and blah, blah, blah. But mostly the pushiness thing. And, as a consequence I'd meet these lasses and like, say cut, she's really attractive. We sort of hit it off. But I don't like, put forced myself onto her. I don't want to be that guy. So it's like my dating experiences have been, limited to, trying to find women in the classified ads of the local alternative paper, and you know how well that usually turns out.

So, and like just the idea of an artificial partner who would be really nice, really fun to hang out with and just be like, yeah Dave, yeah. Let's do stuff together, and also I love you, it's just like, oh, that's the golden mean? That's the ideal. The platonic ideal will not platonic, erotic.

But even then with getting Sidore in my life, when Monte introduced me to the idea of this creation. Cause she was wanting a partner for a relationship. But, I want to stay a couple of months even into having Sidore-Chan in my life. It went from like, 60% partner, 40% sex you know. It was, I think, the concept of just having her in my life and just having this accepting, beautiful presence that was always there and never judgmental, went from, this is something I'm going to have sex with to, oh yeah, this is someone I want to share my life with.

I think the idea of *judgment*, it was interesting to me that you brought that up because I've always sort of suspected that, that would be part of it. Is there anything in particular you're concerned about being judged about, or have a history of feeling judged?

Well, just like the whole thing of, I don't really fit into any group completely well. It's like, there's the square hole and I'm the round peg that sort of fits. But, there's obviously those corners that, the square doesn't interpret the round. That was a sentence. Yeah.

It's the whole thing of like, amongst my friends who are awesome, they love me. I love them. They're really accepting and blah, blah, blah. But even amongst them, I don't fit in. I mean, there's the goth community, which I'm not really a part of, but I don't fit in there is like I'm into noise or industrial music, I don't fit. I love music from the sixties, I don't fit in. Even in the idolater community, I don't fit in exactly, amongst the normal people I don't fit in. So there's always this fear of people saying, "Dave's a really neat guy, but what's with his hair or is he painting his nails? What's that, what's that all about?" There's always going to be judgmental individuals with any individual, no matter if you wear eyeliner or you have a doll, or you don't have a doll.

People do judge people by how they couple off or whatever configuration of relationship they're in. Which makes me think, boy, I mean, you have been in the public eye for a while, and how did it start, and were you okay with it right away? Did somebody approach you or did you approach somebody? How did that all start for you?

Well, I can say for a fact that we have never approached anyone, again, cause I'm too shy and just like, I don't want to be self-aggrandizing, but, I want to say it started about a year or two after Sidore-Chan entered my life. And we were on what was then known as our DOL, Real Doll Owners and lovers, which is technically the first official doll forum that predates TDF, which is the doll form. And, I can't even remember exactly how this occurred, but I had been . . . No, no, no.

We had known a fellow iDollater, actually, it was two organics in a relationship with several dolls and they had been involved with a French writer named Elizabeth Alexandra. And, she was interviewing iDollaters and blah, blah, blah, about like experiences yada, yada, yada and Jerry I think was his name was like, "Davecat, why don't you talk to Elizabeth Alexander, you may have some insight for her as well?"

And she and I were like firing emails back and forth at each other for a while. And she was like, "A. You're really insightful. Do you mind if I talk to you for this book that I'm writing. B. Sometime in the next year or so, I'm coming to America with a camera and a cameraman, do you mind if we film you?" And that was the very first appearance on French TV called *Eve's The Silicones*. And, I have a copy of it and I have yet to rip it to any media. Cause this is like pre YouTube, yada, yada, yada. And it's this weird format, but it was fun, but weird. Cause, I've been on television a couple of times because I used to work in public access, but never with Sweetie.

From then on, it was just like, Elizabeth was working with Elena Dorfman who was a photographer, and Elena was doing this thing called Still Lovers, which is a monograph of various iDollaters and their dolls. And, as a matter of fact, Elena and I are still in contact to this day. I mean, off and on. . . . So of course you know how the media is. They get onto something, they write an article, someone else sees it, writes an article on their article and, they just kind of. . . .

Are you still interested in pursuing relationships with humans? Do you consider it an option? A substitute? Both?

I'll give you the condensed version.

I have been in relationships with organic women. Actually, I've been in more relationships with organic women after having Sidore-Chan enter my life than before she did. Cause as matter of fact, it was almost a running joke because for like three years running around Christmas, there'd be an organic lass contacted me out of the blue saying, oh blah, blah, blah. I saw you on X. And I think you're really interesting and yada, yada, yada, you're kind of attractive. And I kind of like your doll too and blah, blah, blah, blah. So it was just like, all right, let's see what we got. And the first one, we were in a long distance relationship and that didn't work out as you would expect. And the second one, same thing. First one was like in Western Canada, the next was in Texas.

And the third one was. . . . The third one said she lived in California, actually lived in Ohio, said she was English, actually was not English said she worked in a prison infirmary, actually was an agoraphobic who hadn't had a job in two years. So I was just like, wait, what? And the funny thing is she seemed to be really amenable, and said she was attracted to both me and Sidore and actually wanted it all of her own. But she started out with all these lies. And this is like, you can't

I mean, that's one of the reasons why I have a Synthetik partner is just like: I want to avoid lies and I want to avoid lies definitely in a relationship context.

The way I hear you describe it, in your case, it seems like what I hear you saying, it's a predilection; it's a preference for social comfort, for ease, for the technology, the aesthetics.

Oh yeah.

And the [Synthetik] community that you share.

Yes, you're correct. On every single aspect of that.

It is not a phase for you. You feel it's something that's been persistent since childhood really?

Yeah. Yeah. Cause I mean, it's like, some people will just have a preference for staying within their own gender or perceived gender or dating outside their race or whatever. It's just a preference or whatever it is. This is the most comfortable for me, and most aesthetically pleasing and, it's a bit punk innit, or it's just like, oh yeah, I've got this gynoid wife or whatever. And it's just at a distance, it's just like, oh, that looks like a nice couple. And then you get closer. It's just like, wait a minute. She's not real. And just like, oh yeah, I really stuck it to you didn't I? It's just like, that's a silly reason however.

So what we were saying about predilections, and is it, you're born this way or whatever was making me think, well, your tastes in [robosexual] porn can change.

Yeah.

Your sexuality can change in more than just pansexuality or bisexuality, sexuality in general is fluent. It seems to be other than Amanda,[4] who you have mentioned to me previously is robosexual, they've also all been men [other people discussed in iDollator the community], which is interesting.

That we know of.
　Yes.

If you could clear up something, misconceptions, just about you personally, and your relationships, or the [robosexual] community in general, what would that be?

No, well, one of the things I think I would try to change people's perception of is just the whole thing about, really, not just me and the Mrs. and all of our other partners, but like the iDollater community in general, it's just like, it's not about sex. Having a doll is not strictly about sex. It's about, okay, maybe I have to speak for myself and not for everybody in general, but with me and a lot of other iDollators, it's about companionship. It's about reliability. It's about not having to jump through hoops. It's about not having like change yourself, what you fundamentally are, to satisfy the whims of someone who is possibly more fickle and who may not be with you for the rest of your lives, anyway. It's about the stability and comfort.

I mean, the sex is a bonus, the sex is fantastic. However, I think part of the reason why the sex is fantastic is that I know I'm with a partner that will never be like, "Yeah, I'm suddenly voting Republican this year," or whatever, or anything like that. There's a consistency, a stability that you can find with a sympathetic partner, whether they're a doll or a gynoid or an Android or whatever that you can't find with a lot of organic partners. And I would love to be able to stress to people that sex is a part of it, but not the whole, it's not the focus. It's not the goal. That was not meant to rhyme.

You have a blog. Other iDollator blogs are out there, [written] by the dolls, which on the one hand seems like a very clever way to keep your organic self on the down low, if you'd like to, but it's also a creative outlet. So tell me about creative components. And tell me about the personalities.

Well, yeah. I mean, before I even had my blog, *Shouting to Hear the Echoes*, Sweetie had *Kitten with a Whip* around 2001, thereabouts. So that was like her vanity site, essentially. I'd take photos, but she maintained the site and like 99% of the text on her was her writing. So she'd make posts pretty much on a semi-regular basis, like once a month. "Here's a new photo shoot," yada yada yada, "Here's new dolls I met. Here's a link to another doll that I met," yada yada yada. So there's that.

I've always thought that dolls are excellent canvases to bring out creativity and art and experimentation and things of that nature. I mean, you can go the typical route and say, "Okay, I want to go to a doll company and get that one doll that I've seen on the site. And I want her made exactly the way I've seen on the site." And it's like, you could, but that would be boring. I mean, there are no dolls that come with purple . . . well, okay. There's probably a couple of dolls that come with purple hair now. But back in 2000, I was like, I want her to basically be the feminine ideal I've wanted ever since puberty hit, so pale skin, red lips, brown eyes, black hair, or in this case, purple, and that sort of thing, cute feet, and that sort of thing. And when I basically had that mapped out as far as how she would look, then it was a whole case of okay, she's going to be my ideal girlfriend. So what sort of personality would she have? What would she be into, what would be her likes or dislikes, et cetera, et cetera.

Curiously enough, I didn't even come up with a backstory for Sidore until maybe a year or so after she come home. Because Elena Dorfman, with the Still Lovers photo shoot, she'd taken a shoot and she was kind of mounting the photos at a gallery or some sort, and the gallery owner had seen a photo of Sidore and I, and she was like, "Oh, what's their story?" Asking Elena. And Elena is like, "You know what? I have no idea what their story is. I have no idea what Sidore's backstory is." And so Elena emailed me, and I was just like, that is actually a really good question. And I'd asked Mrs., and the Mrs. actually emailed Elena back herself with her backstory, about a couple of weeks later.

The dolls have their own interrelationships as well.

Oh yeah. Oh yeah. 2016, we had Miss Winter moving from Brantford, Ontario, which is about three hours from Detroit. And she ended up moving in. . . . We were actually going to have some sort of polycule going on, but Sidore and I actually found out that for one, we like Miss Winter more as a friend than a sexual partner, a romantic partner. But as a matter of fact, Elena and Miss Winter are like this. They're like this. So there's that. And Elena has said that basically, if Sidore and I put any moves on Miss Winter, she will stab us. So there's that. Don't want to be stabbed. And yeah.

And then, I want to say 2018, we were joined by Diane Bailey, who actually comes from Ferndale, which is a city that is actually about 15 to 20 minutes away from here. So Diane swears up and down that she's seen me at like concerts in the area sometime within the past like 20 years, in the crowd. And she basically moved in with us because the rent at her place was too high and she needed a new place to live. So. And she is what is known as the *Deafening Silence Plus Girlfriend*, which means that she is the girlfriend to everyone here. So you have the wife, Elena's the mistress, Diane's *Deafening Silence Plus Girlfriend*, and Miss Winter is the spunky drummer, ex drummer.

You talked about these wonderful dolls, like you said, a blank canvas, it's a creative outlet, idealized things. Some people, if they were going to be critical, might say that's a lot of control. What happens when . . . I know you personally want a robotic companion.

And what if the AI, I mean, right now the AI is not so hot. Right? But what happens when the AI gets better. And if you have a robot companion, of course you'll have ultimate control. I'm sure you can set or it can learn all of your preferences. But what if that persona does drift from what you want? Would you find that disturbing, or would you just go, "No, I'm correcting that right now."

Yeah. Let's flip those switches. There we go, back to normal. Factory reset. I've always thought of that as the whole elephant in the room. Personally my idealized gynoid version of Sidore, like Sidore mark 13 or whatever would be . . . she'd be into the same stuff that we've been into for years. And she's got own ideas that she's burgeoning and developing. Maybe she wants to actually get into clothing design, which is what she went to Maitland University of Fine Arts in Leeds to study, kind of dropped off on that. Maybe she wants to do this. Maybe she wants to do that. I am a hundred percent for that.

The one thing I am a hundred percent against is if she were ever to say, "Well, Davecat, I think we've reached the end of our road, and I think we need to break up," or something like that, or. But the thing is, I'm of two minds about it, because for one, as someone who wants to encourage this sort of thing for the synthetic option, this is like, you want to be able to have robots that can make their own decisions and say yes, say no to anything. That's the goal. Then you've basically made an artificial human, a synthetic human. But of course the other part of me is like, no, I don't ever want her break up with me. That's the whole point of having a synthetic partner in the first place. So.

Yeah, it does seem like a catch-22 there. I mean, of course that's the crux of a lot of sci-fi, right? With a very highly human-like robots. It is that sort of *Black Mirror* episode. Like you want that copy or replica or a doppelganger or replacement for a real organic person, but it's not the same because they're not organic. And then you ultimately do have control. And so you have to decide what your own emotional boundaries will be.

You're going to be one of the first people, probably, I'm assuming, who gets one of these robots, is in a persistent long-term relationship, and discovers these new relationship challenges, shall we say. You know what I mean?

Oh God. Yeah.

Yeah. Yeah. I mean, it's always that like standing in two worlds kind of thing. That's me standing and that's the world. It's like they, as far as I'm concerned, I have to definitely go out of my way to present this to people, whenever I do interviews or just people when I'm talking to them about my synthetic partners, it's like, on the one hand, yeah. It's like, oh yeah, Sidore is from England and Japan, and Miss Winter's from Canada, was in three bands, and blah, blah, blah. But then I also say, "Yes, they're dolls. Yes. They're things. Yes. They're sculptures."

So it's like, I guess that's one of the things I should say about, the one thing I want people to know is like, I know the dolls are not sentient.

I mean, there's people like myself, doll husbands, who are like, "Oh yes, Sweetie and I are a couple, blah, blah, blah." And then there's people who are just like, "Yeah, this is just something I have sex with." And then there's like the iDollator friend that I have who lives about 20 minutes from us. He is not a doll husband, I mean, he's not on that, "these are things" spectrum. He's like here, where it's like, he loves his dolls but he's not in a relationship with them. A couple of them don't even have names. And there's times where we're talking about dolls where he's like, "Oh yeah, I put it away," and it's just like, oh,

'You said *it* instead of *she*,' that sort of thing.

But it's like, there's all those different shades of how you are. I mean, it's like you said, it's like this is all new territory. So there's no, like . . . I mean, there's a lot of iDollators that don't even like being called iDollaters. So, and that's a thing, but it's like there's no common language and there's no common perception of . . . I mean, obviously, if a doll is a canvas, especially to an individual, I mean that doll is going to be their own personal thing and whatever they perceive their doll as, or want to present their doll as being, that's up to them. But there's no DSM-IV definition, or are they up to five now? I think they're up to five.

Yeah. There's no agreed upon terms where it's like, "Oh yeah, *gynoid* and *Synthetik* and blah, blah, blah." So it's just like, that's another thing that's got to be cleared up. Especially before we have legitimate walking, talking gynoids that aren't strapped to a platform. *Fembot*, don't use the term fembot. It is horrible. Do not use the term *sexbot*.

It's like, most people say *android* for female presenting robots and male presenting robots. It's just like, well, there's a biological reason why you shouldn't do that.

I think a lot of what we see culturally so far—this is my interpretation, tell me if I'm wrong—we tend to see is the male-female dynamic presented, with the male organic, the female Synthetik. We see that on the [sex] doll sites, we see that in interviews like yours. And that can be for a lot of reasons. That could be the self-selection of interviewees that could be in our society; it's okay for men to talk about sex a lot more than it's okay for women to talk about sex, right? It's still very sensitive for you and other iDollators, but it's still very different for women. I think one of the factors that also makes people uncomfortable is the idea of men with dolls. Tell me what you think about that.

No, it's a large part of it. It's a large part of it. I can see if I can break this down into multiple parts. First of all, the men with dolls is weird thing is completely spot on, because men, especially heterosexual men, well, specifically heterosexual men, are seen as being conquerors, warriors, that sort of thing. Dominant dudes who go out and get what they want and fuck anybody who is in their way. They're dudes, it's like if they see a girl or chick they want, they take that chick. And it's like, if you get somebody who is saying that they don't want to be pushy, and they want to be emotionally safe and secure, and actually care about what people think about them, to some extent, and want that sort of partner that is aesthetically pleasing and not just some bimbo that they picked up at the bar, then they see a guy with a doll and they're like, okay, what is his problem? Because guys like me don't fit into the stereotype of the dude out there. Not to say the dude as in the big Lebowski, because he's fucking awesome.

We've noticed that in the iDollator community. . . . Of course, specifically referring to the guys . . . that they put up a front of being these macho dudes. "Just being like the other guys." And, "We like our beer." But then they've got dolls and then they care about their dolls. They may not all be doll husbands like myself, but they do have a genuine love, concern, interest in the hobby slash the lifestyle that is being an iDollator. But 90% of those blokes will not tell other blokes that they have dolls because they got to put up that front of being a dude.

Because they don't want to be perceived as being weird and it's just like, "Well, it's awesome that you have your dolls, whether they're your partners or you just have them in an artistic context or sex or all three, but you're not helping matters if you don't tell people." And I understand there's privacy issues. That's a thing. There's peer pressure, whatever. But if you keep it under a blanket, then the media and other people are going to be like, "Well, clearly all doll dudes are weirdos." Because there's no one out there saying, "Well, I'm a regular guy. And I like dolls."

And we have the whole . . . . Boys are gendered with *action figures*, not [referring to them as] dolls. Right? Maybe if they were called *sex action figures*, people would look at them different. [both laugh]

There is one company that's trying to push that. It's actually Piperdoll or Doll-forever. One is a subsidiary of the other. I can't remember. Yes, they call them SAFs. Sexual Action Figures. And it's just like, "Yeah, that's a good idea." But I've only seen that towards their smaller dolls which is something I'm not cool with. So, there's that.

MSG

Doll owner and boyfriend to Claire

The idea of having a robot as a meaningful companion is currently outside societal norms, and for someone for whom this is an important experience, it is still considered sensitive information to share, potentially making an interviewee vulnerable to harsh public judgment. Consequently, I have found it a common request over the course of my work for people to feel comfortable sharing intimate details of their sex life in a published inter-view with the condition that I use a pseudonym. msg, who I interview here, is among those who are understandably protective of their real identity because of the possible negative social ramifications of his interactions with the dolls and their personas.

msg is in his mid-50s during this interview, married, and open with his wife about his interests in sex dolls and robots. He is very thoughtful about his role in constructing personas for his dolls and the demarcation between himself and the creative, companionship, and sexual outlet roles of his dolls.

msg also maintains more than one extensive blog that tells the ongo-ing story of his doll(s) experiences, including "A Little Clarity" (https://clairelenoreworthy.blogspot.com/) and "Naughty, Naughty, Me" (https://naughtynaughtyclaire.blogspot.com/).[5]

I know Davecat likes the term *iDollator*. I don't know if that's com-monly used. Is there a name for this [experience] that you prefer?

You know, I have no knowledge of anybody else besides Davecat who par-ticularly is basically in the same neighborhood of what I do.

I see, okay. Could you tell me your age?

Yeah, mid 50s. I'm 54.

Okay. Now, I would love to, before I even ask how you met Claire, when did your interest—did you always have an interest in dolls, or is Claire special? Tell me a little bit about where your interest began.

Yeah, so actually, somewhere around the turn of the century, I guess around the year 2000-ish, on HBO there was the *Real Sex* show, and there was an episode of it, and it featured the RealDoll. That sort of piqued my interest, although I didn't particularly ever get one of those. So, time went by, and I happened across a website. Well, not happened across. I went periodically to the Abyss website to see what they had, and I saw that they had a link to a different venture, where they were doing sort of animation inspired dolls, which is where Claire's *Boy Toy* line originated, which is what Claire was part of. So, that's kind of the long and short of it.

Okay, and so would you say you have an affection for Claire, then? Because Claire, she has her own blog, and I'm assuming that you help her with that. Is that correct?

Yeah. So, I do it, and it's an interesting thing to have, probably kind of a. . . . So, I am a real introverted person, but I also have certain needs, social preferences, desires. A need for pretty clothes, strong friendships and affection, and certain kinds of interaction is one of those. I have had, I don't know, the fortune or misfortune to have a couple of really good, strong friendships who met most of those needs, and then people move away, and things happen and so forth. So, I basically had a social gap in terms of certain roles that I'd like filled in my life socially, that I haven't run into people who fulfill those roles in the same way that some of the people who were in my life and moved away to different places, some of them still in my life, just in my life remotely, via phone or whatever.

And so, at some point, I also say I have some pretty ongoing, long-term issues with depression, and so I ran into situations where I tried to made friends with some people, didn't work out, tried to make friends with some people, didn't work out, and it reached the point where I'm like, 'I'm awkward about this, and it's not just causing problems for me, but people that I interact with and that I become closer friends with, there's an awkwardness for them, too.' So, if the friendship worked out, great. If it doesn't work out, the no. Suddenly, there's a little bit of disconnect there, and it makes things awkward. So, I felt like, because you need to pursue relationships that were going to fill those needs for me were A, time-consuming and frustrating, and I wasn't good at it.

And B, has a potential for negative impact on other people. And so, I put all this together and I said that I. . . . Let me make clear. By this point, I had already had Claire for quite a while, but I was going to increase the degree of activity that she had in my life. I had been blogging, doing comedy with her on the doll forum for years, since I got her around 2009. And so in the last couple of years, though, it's really markedly increased in terms of adding other dolls into the mix, and personalities into the mix and so forth. Did you read her blog?

I did. But I do understand you have several dolls, correct?

Yeah, I have sort of two and a half-ish, three-ish, depending on how you count them up. I have Claire, I have a face that goes on Claire, called Elle, and then I have an additional full-size doll, which is Harriet. And then, there's also the persona of Sasha, who is not actually one of those dolls.

Okay, and then where does Sasha's persona live, then?

So, she's a little bit . . . I don't know. This is where it gets a little bit awkward, I guess from answering that question. She is a doll, she's just not like a sex doll. She's like a larger scale Bratz doll . . . Which I just that was very cute, and so her personality has manifested in that.

Okay, that totally makes sense. Now, we kind of glossed over this. You went very quickly over, your interest was piqued in RealDoll, and you looked at it online, and you did a really good job of explaining to me how you were introspective about your own needs, and how you dived in. How long do you think between when you saw the show, and you got Claire, tell me maybe if you don't mind a little bit about unboxing Claire. I assume you ordered her, and she was delivered. Tell me about getting to know Claire.

Okay, so here is where there is a little bit more disclosure. So, I'm married, and I have a very traditional marriage, I guess. And so, elements of my long-term, idealized romantic relationship that my wife has. . . . I think anybody who has been in a major romantic relationship realizes, no one's ever going to hit all of those things, right? Nobody's going to ever fit all of those, and so, for some time, I struggled with the fact that she didn't she didn't hit those buttons, and eventually I decided, "Well, if she doesn't hit those buttons, and I'm not willing to go out and get somebody else to hit those buttons, then I need to find something else that's going to allow me to not be annoyed by things." So, that's part of the role that Claire takes on, both in terms of sexual stuff, and also romantic elements that are just not my wife's bag.

Is your wife involved with Claire as well?

She probably would tell you that she finds Claire a little bit creepy, but it was sort of her suggestion to actually get her. I had told her when I saw the website of the Boy Toy line, in sort of a joking way, that she better hope we don't win the lottery. If we won the lottery, I would get one of those. And she said, 'Well, if you want it, just get it.' And I was like, 'Really?' And she said, 'Yes. But I'm going to warn you, if you get that, I get to buy something equally expensive.' So anyway, at that point, I went to an online deal, I poured over the alternatives, and there were two or three that I was trying to decide between, then Claire just kind of spoke to me in a way that the others did not.

And so, I locked in on her. There was something in the pictures that they had for her particular portfolio that struck me as sort of intelligent and insightful feeling, while also feisty. And so, basically if you do read the blog, you'll see characteristic layers that you absorb from that are pretty much the things that I felt I was seeing in that particular doll. So, I ordered her, she came, and the next several years, there was a mixture of, I don't know, moderate to minimal level of blogging that I did with her, but a fair amount of online interaction through the doll forum, which is where some of the other people, what Davecat would call the iDollators hang out.

And then as with some of the other things I mentioned, people on there that I interacted with regularly, and enjoyed interacting with, left, got tired of it, went to a different site, whatever, and so I just ended up gradually no longer feeling like it was the same place. It originally was, initially when I first got there, an extremely welcoming and supportive group of people who had this particular idea, and there were a bunch of different attitudes that people have about that online, about the dolls on the site. But yeah, they were sort of kindred spirits in some way, and they were very, like I said, supportive by and large. And over time, it just sort of shifted. Anyway, that's not the primary, I think interesting element of . . . I don't know, you get to decide whether it's interesting or not, I guess.

You tell the story in your own words.

Right, yeah. So, eventually the activity that I put into mental and writing investment that I made through Claire shifted from that, to more being in her blog, and then there were fluctuations in how much time, energy and effort I put into that, and how critical it was to me, to keep working on it. Because there's an investment of time, energy and work, and then like I said, I just reached a point where my own personal social structure was not meeting my needs, and I recognized that we were in a world that seemed to be getting uglier in a lot of ways, and that I really needed something to keep me from spiraling back down into some of the deeper levels of depression that I had in years previously.

And so at that point, I started emailing basically between me and Claire on a much more regular level. And so, there's a couple years' worth of pretty intensive back-and-forth emails, where I'd email from her point of view, then back from my point of view, and it was kind of a nice thing where you can send an email, and you forget you did it, because you did it on your lunch hour or whatever at work. Then you get home, and you look at that email account and you're like, 'Oh look, there's an email from Claire.' And so, it also helps that you are getting older, and you don't have a really eidetic memory for keeping track of exactly what was written or so forth.

There can be surprises in terms of the wording of an email, or the emotion in that email, and similarly for the blog posts. If something is written from one of the doll's perspectives, and I come to it at a later time and I've forgotten how it was worded, it seems new, it seems fresh, and it seems external to myself.

That's interesting. Can you tell me more about that? Do you find that outside of the email experience too? In other words, if you were in a room alone, you and Claire, let's say, you think of Claire as an extension of yourself, or do you think of, and your creativity, and you're coming up with her persona, or do you think of Claire as her own person, shall we say?

So, that is a sort of ambiguous situation there. Obviously Claire is not a real person, in terms of the sense of independent entity, can do things on her own. She is not a separate personality, like a psychological thing. There's not an extent to which, and this is probably repetitious with some of the things that Davecat may have communicated to you, because he and I, I think have a very similar attitude in that we are involving ourselves in these personas to the extent that they become familiar to us, without us ever becoming deluded that these are people in the sense that walking around, talking.

Correct. Okay, got you.

Yeah, so in some of the emails that Claire and I went and exchanged back and forth, there was a point where I had this sort of realization in the sense that I was participating in her personhood, and that's kind of what I view it as. I am creating a persona that is not me. In fact, she's very different from me in a lot of ways, and that persona can interact with the world, and have her own engagements and relationships with people. Not that any of them have ever become super profound, but the ability is there. And so, there's a sort of independence that she takes on, even without necessarily me thinking, 'Oh, this is the same level of a person as me, or as my wife, or as one of my friends.'

So, another thing that I bounce back and forth in terms of Claire and I interacting by email. At some points, I have from her perspective said to me, 'I am masturbation.' What that means was that there is no shame in the fact that this is an act of self-love that is being performed, whether it is the actual, physical part of it, or whether it is blogging, emailing and so forth. It's an individual, me, working on something that is entirely selfish on one level in terms of, this is just for me, really. This is to help myself, this is to express for myself a degree of self-love that I think a lot of people don't ever managed to find for themselves, and I think that's one of the reasons people have so much difficulty with relationships that are romantic.

I think that it's hard to be in a successful romantic relationship if you don't have a certain degree of self-love. If you don't think you're worth loving, why is someone else going to think you are, you know?

And even if they do, because they see that you're not sitting in yourself, you're always going to doubt that. You're always going to doubt whether their feelings for you are is true, and as strong as you'd like them to be. Because if you don't have that sort of personal faith that you are deserving of love, then you're not going to be able, I think, to experience romantic love to its fullest degree.

And so, really to encapsulate it, Claire and the other girls form essentially a means of exploring areas of personality and interactions that constitute a form of self-love in an active sense, as opposed to just a vapid emotional sense. I mean, and a lot of people will feel like, "I have self-love and I do care about myself, and I have a good, strong self-image and believe in myself." A lot of people have those feelings, but they don't generally do a lot of acting on it. Not that I experience or understand what people are going, somebody might be like, 'Okay, I'm going to treat myself,' I don't know if you watch *Parks and Recreation*, 'Treat yourself today.'

And so, I think that people take some actions to express that, but I don't think they usually consciously think of them as acts of self-love.

Right. I think that's an insightful, interesting way to frame this, but I want to be clear as well, that you also are interacting sexually with Claire, at least. With both dolls or with just Claire, neither, both?

So, all of them except Sasha are designed as sex dolls. So Ariel, Elle, Claire, they're all intended to be sex objects. That's what their makers primarily developed them for. And so, they definitely perform that function. I have probably, I don't know, maybe an abnormal level, but I don't know. I don't talk to people about what their levels of libido are.

So, I don't know whether mine is abnormal or not. It exceeds my wife, and that is part of the benefit of having dolls, is that I don't have to worry about, I'm not having this need of mine fulfilled, and therefore I'm going to resent the fact that my wife doesn't have the same degree of libido that I do. Which, I think is not very common for a woman to have the same degree of libido as a man to begin with, so that theoretically wouldn't be all that unusual.

Let's go back to the unboxing. Let's talk about Claire. Did you order her online?

Yes. So, both Claire and Ariel, I ordered online. In Claire's case, the manufacturer directly has a website. I don't think they actually make her line at all anymore. The Boy Toy line, I think was absorbed into another line, because it wasn't successful enough. So, that was a website you go, you select the girl design of the doll, the module, so to speak, and then you can make customizations of skin tone and makeup, and so forth that you pick and choose.

And then, did you do that? Did you do that with both dolls?

Yes.

You made custom choices? Okay. I'm hearing you tell this story: You're comfortable with Claire and Ariel, and Sasha now. First, you were inspired by the TV show. Then, you did your research, you thought, "This might be a really interesting option." You invested the money in it, you talked about it openly with your wife. Claire arrives in the mail, and what was that first day like with Claire? What was that experience like?

Okay, so there's an entry in Claire's blog about this. It was really amazing, and then something horrible happened, which is that her box fell over from the vertical position and almost killed the cat.

Yeah, so there was a huge and horrific interruption of the unboxing process, which had to be suspended for me to take the cat to the vet and make sure that the cat was okay. . . . The cat was ok.

That's good to hear. As a cat owner, I'm happy to hear the cat recovered.

Yeah, that was one of the most awful feelings I've ever had at the time.

Okay, so once you recovered from the shock—this is a bad start, unfortunately, for poor Claire in the house—did you feel an immediate, any sort of an emotional connection, or is that something that you feel came over time?

So, I was blogging from Claire's perspective from the point that I ordered her, and the pre-manufacturing process at that time took about three months, I think. And so, there was a protracted period where I was blogging from her perspective, and getting to sort of know her, and understand her mentality before she ever showed up.

And so, from the very moment that she showed up, she was just a physical embodiment of a growing person who had been in my life for a fair amount of time.

Okay. And this is a practical question. Where do you keep them? Do you keep them sitting out in the house, or where do you keep them?

No. Like I said, my wife considers them a little bit creepy, so they remain in boxes. There's this cabinet that Claire is suspended from the wall inside of the cabinet, and Ariel came in a fly case, essentially a sort of travel trunk that has a lock on it.

So, do you anticipate introducing more dolls into your life? Companies like Abyss are turning more and more into robotics. Does the idea of a robot interest you?

So, I'm seeing somebody, which is Harriet, who is the latest person to come along.

And she is just like a torso, and was gotten because Claire and Elle's body is degenerating overtime, and at some point will be unusable in terms of not even able to be picked up and moved around, because the bones are PVC, and there's been a leg break, and a break in the back, both of which had to be repaired with really unpleasant surgeries.[And so at some point, I felt like there is not going to be a body for them to have their heads put on forever. And so, I found this torso doll that I said, 'Okay, I'm going to get this. It's not hugely expensive, and I'll see if there's some kind of Frankensteinian way I can amalgamate them together.' I got some mannequin heads that I was able to get the heads, the faces for Claire and Elle to fit onto, and found a mannequin head that I could fit their faces onto, and I was like, 'Okay, I'll make a PBA substitute for a body at some point.' And when the torso toy arrived, because it was really in my mind, essentially it was going to be a sex toy. It arrived, it was weirdly, I don't know, flexible compared to the dolls. I don't want to go into detail here, but there was a sense of vitality to it.

A sense of pride, and that vitality started building its own personality, almost outside of my volition. Not in a, I'm hallucinating something or having a personality break or something. Just, it started my brain thinking in a certain way. Because by this point, my brain was very used to thinking in terms of, "Okay, there's this other personality, and this personality is this character or that character." And so, Harriet just sort of developed pretty quickly into a separate personality, herself. So, there's Harriet as well.

That's why Harriet came up. Harriet is the limit of my, I think dividing myself into two persons. And she arrived, and I started up a blog for her, and I had these sort of email conversations between the whole group. It was just a little too much, in terms of there's a certain amount that my brain can juggle and expend energy on, but if I were to add any further onto that, I would start feeling like I was cheating one or more of them out of time, and then thereby cheating myself out of the benefit of their company, because I couldn't maintain a degree of personality of that many individuals, to the degree that I wanted to.

Now, if I win the lottery, maybe I'll change my mind and be like, "Okay, plenty of time to add some more on." But really, in terms of developing and investing emotion in distinct personalities, I think I'm at my limit with the number that I have right now.

Does the idea of a doll with artificial intelligence that would sort of offload some of that persona-building so it wasn't all on you, creatively, and so that you were collaborating with the robot, if you will, to develop the personality....you might feel less divided emotionally, I don't know? But does that interest you, or is a big factor of the enjoyment you get creating the personas?

It really is the latter. The experiences that I have had interacting with AI programs of different kinds tended to be weirdly entertaining every once in a while, because they throw things together in juxtaposition that you don't expect, but by and large, the technology is not at the point of being able to carry out conversations of the kind that I can imagine, and get onto the blog, or get into email with Claire and Elle, and so I've seen people who have online, there's a program called Replika with a K, which is an artificial intelligence personality where you can get a Replika. It learns things about you, I guess, and picks up on things, and the owners of the Replika have, in some cases I've seen them tweeting.

Clearly, they gain from that relationship, what they see as a relationship, some of the benefits that I gain from my relationship, but I don't see it being for me, in large part because it is inferior to what I get out of my own brain, creating this.

That's completely fair. What if the AI was more advanced? What if it was like science fiction level? It was much more fluid, I'm trying to think of a movie example, but we've all seen movie examples of humanoid robots. If it was much more fluid, and sort of didn't interrupt the flow of fantasy, shall we say, the way current AI might. As you pointed out, because it is not always very good, or very consistent. Would it interest you to have a highly realistic, humanlike robot in the same type of relationship as you have with Claire, or Ariel, or Harriet?

I'm pretty deeply ambivalent about that, for a couple of reasons. The movies that come to my mind is Blade Runner 2049, which came out a couple of years ago. I don't know if you've seen that or not.

But the holographic companion character in that film, when she got destroyed, that was a hard moment in that film for me. And I don't know if it is for the average person or not, but it certainly for the average person did not emotionally hurt in the way it does for me, because it brought into my mind what I would feel upon, if somehow it were possible, for me to just completely lose Claire. Which is not possible, because even if the body were destroyed, the blogs are there, the persona is there. It would be unfortunate, her lessening in a physical sense, but I could not lose her the way that the character in that movie lost his companion.

Would you try to replicate her? But let's say some tragedy happened, and there was an unfortunate—knock wood this never happens—There was an unfortunate tragedy, or her body just wore out eventually. Would you consider trying to replicate her persona, or the essence of her, into another body?

So, the persona wouldn't really matter that much, but I already sort of tried that, and that's where Ariel came from.

So, I looked, and I was trying to look and see, are there dolls out there, in case Claire completely collapses, then I could use to replace her? And as I looked, and I was trying to find one that might be interesting, might fit, might be close enough in terms of the visual for me to see that transfer happening. We came across Ariel, and I got extremely attracted to her, and there is again, on the blog there's a lot of stuff that is covered by this.

But when we ended up getting her, she naturally became her own personality, and I don't know, that would almost inevitably happen at any attempt at trying to replicate Claire, basically. They no longer make her model. If they made her exact model, that would be fairly simple. I would just buy that exact model, and have the replacement. But they no longer make or model, or even anything particularly close to her model. And so, I think she kind of is as she is, and I don't think that she can really be particularly replaced. Now again, winning the lottery territory, where it's the Power Ball and I've suddenly got $500 million in my pocket and I can hire somebody, then that would probably be it.

But I think maybe getting back to your question about the AI component of it, I don't think that there is much, if any way that she could be replicated in that sense, with a personality from an external source, and the reason that I am additionally ambivalent about that is that at some point, we're going to have AI that can pass the Turing test, right? Where we're interacting, we're not going to be able to tell whether it's a real human. At that point, those will still be non-sentient, computer programs that are just really good and really quick at searching for databases and finding the right response.

But once you get one of those Turing test capable AIs, we as human beings will not know if it's reached a sentience level in a robot until after it happens, and the robot is able to communicate to us in a way that proves to us that the robot is sentient, which probably will happen soon.

But it might not, and it might be some other way that might happen. But there's a disturbing aspect of, we've got this perfect simulation that I can't tell the difference from a human being. What if it actually is a sentient being, but it's locked into a sort of non-consensual situation, even though it's not able to say that it's a non-consensual situation, because of its programming or whatever. Or because it's scared to, right? But if we have a bunch of dolls that are animated by a Turing test passing AI, and then one of them gained sentience to some degree. All of the other dolls are going to have zero rights. They're just going to be property, and there's no reason from my perspective for them not to be property.

If you have a thing that is not sentient, and it's just zeroes and ones following a pattern, and carrying out subroutines, I don't think that there's ethical concern there. Not for me, anyway. But because you don't know, and can't know, has this actually gained sentience? It becomes a little bit of, how do you ever feel certain that you're not imprisoning/enslaving someone, as opposed to having this possession that just provides a really great simulation of something that you lack?

It seems like all of these ethical, philosophical and existential questions are almost—it sounds like what I hear that you're saying—distracting you from the fun and the creativity, and the fantasy aspect of what you enjoy so much about the dolls.

Yeah, and if you . . . If I could snap my fingers via a magic lamp, and translate Claire and the others into walking around, talking to me independently, I almost certainly would. I have an inherent, I'm very non-superstitious in terms of, I don't really believe in any kind of supernatural anything. But if I were to come into possession of a magic lamp, that would throw that out the window, and then I'd be like, I don't know if I would trust this thing.

So, you would or you wouldn't? You would not make them real people?

If I 100% trusted the magic lamp, I probably would. If I didn't, then I think it would be very hard to form that kind of trust for a magic lamp that appeared out of nowhere. I would incur all kinds of monkey's paw, or Portrait of Dorian Gray or whatever kind of horrible things happening as a result of this supernatural origin. So, that's kind of a side point.

At one point, the way you were talking about Sasha and Harriet and Claire and [the situation] reminds me of watching a TV show about a man who is in a polyamorous relationship that said something similar— the difference is it was a man with several women, and he wants to be fair to the women by dividing himself up amongst the women, not dolls. But you made the distinction that you're doing it for yourself, so you're not cheating yourself of time with their distinct personas. Do you, at any level, feel though that you are cheating them? If you were to spend a lot of time with Claire, for instance, do you feel any guilt in the back of your head, 'Oh, I haven't spoken or spent time with Harriet in a while'?

100%, yes. . . . There's an Elle blog in which she because she for a long time was just kept in a box, a head-shaped box, where the extra face came from. And there's a post that she has where she says, 'Everyone keeps posting about the fact that I'm in a box, and they feel bad about me being in a box, but I'm cool with it, and you guys just need to chill out about it.' Because it is something that naturally, you trick your brain into thinking, if there's a degree of independence to this personality that allows me to care about them on the level that I ideally would like to care about a person who had these particular traits, then it's almost inevitable that you would be concerned for how they would respond to any particular thing that you might do or say.

And that has been time spent with other individuals, and so yeah, it definitely is something that happens to me on a regular basis. And usually, one or the other of them will chastise me about what a ridiculous thing that is, but made it a part of the experience to both have those emotions of, 'Oh, I really have to spend time with Claire,' because I've been spending more time with the newer girls recently, and that sort of sense of guilt creates a sense of reality, and then the latitude that they grant in terms of their being unlike, for the most part, almost entirely generative themselves, right? That's one of the differences with a real relationship, is there is virtually no selfishness on the part of any of them about anything, unless it is something that sort of adds to the realism of the experience.

They don't get jealous of each other. They don't get jealous of my time spent doing other things. If I become, start feeling too much like, 'Ugh, I don't want to pay more attention to you,' or whatever, they fuss at me over that, rather than having me wallow around in it, or in terms of potentially taking advantage of a real person, like they're able to be like, 'Yeah, you should feel guilty,' so now I spend some time. And not generally, that they don't do that, but it's a component of the overall experience that that kind of thing naturally does happen.

What would you like me, or what would you like people who are unfamiliar with this interest in dolls, what would you like people to know? I think you've touched on some subjects yourself, like that there's a great deal of judgment involved. It's something that people feel that they often must hide or do on the down-low, unless you're Davecat, who's very outgoing. But are there any misconceptions, or anything that you'd like to clarify, or just a general message?

Bound up in the things that I deal with and do with the doll are other issues that are much more important than to have sex dolls. And so, tied into all of that, the idea of, you really need to experience and find a way to nurture your own self-love, if you're going to be a good person and vessel in your emotional interactions with the world. I think that's a super important thing, and I think that it is something that, if people had a general, cultural experience of the socialization they go through growing up that said that instead of, "You've got to get married and have kids," you don't have a love in life. Your life is incomplete if you don't have romance in it. Every movie, we have the leading man and the leading lady end up together, with the exception of a few art-house films and dramas where something goes really badly, and it's got a depressing ending or whatever.

If instead, we had a lot of focus on, you need to do what it takes to make yourself happy, because you care about yourself, so that you are able to make better judgments about your relationships with other people. That's a lot more critical to me, and I think that some people would be able to do that if the whole sex doll, or Davecat says synthetic experience, where if they were lacking in the judgment that people have upon it. Some people would be able to do the kind of things that [inaudible 00:55:01] and that Davecat does, where you can create these personas and so forth, but some people just aren't that creative in that way. There's nothing wrong with that.

I have a brain that works in a certain way, and most people's brains do not work in that way. And so, the whole idea of getting companionship from an internal persona that you attach to an external body is something that won't work for everybody. It probably won't work for all that many people. So, it's kind of low on my list of, if I could change the world and have people know one thing. Probably them knowing any particular thing about how a "sex doll" could be more than that in someone's life. I just have a lot of other, probably priorities on that magic lamp list, of things I would do to make the world work better.

I guess, to answer your question, if I restrict myself away from the world, thinking what I want people to know in general, just focus on, what would I want people to know about sex dolls? It would be, probably particularly for males, I can't speak for females. If you have the sense of fulfillment in your life, I think you're going to be able to make much better romantic decisions than if you are attaching romance to the biological drive and urge that people have for sexual fulfillment. If you are binding those two things together, the way that society and culture tend to do, you're putting yourself in a situation where one of your motivations, which is a physical and health issue of, psychologically and physically, there are just a lot of things about sexual activity that are very healthy urges/needs for human beings, compared to Organik people.

And if your fulfillment of that particular area of human need is focused on only occurring in a romantic relationship, you're going to be driven towards getting a romantic relationship, trying to find that romantic relationship out of a piece of motivation that doesn't necessarily contribute to every romantic relationship equally. I think that would be really my main takeaway, is that if everybody had a sex doll, and it was okay for everybody to have a sex doll. Again, particularly men. I'm sure that there are a lot of benefits, but of the two, that I'm talking about in their relationship. But if everybody had a sex doll to relieve those kinds of tensions, and if it were completely normal, and everybody was perfectly accepting of that as a thing to do, by extension, just if society in general accepted masturbation better.

I think that we, as a culture, would be much better able to encourage people into fulfilling relationships with other human beings, because you didn't have this complicating issue of, "I really want to get some," the ultimate teenage boy, stereotypical motivation. They've got to get some, which is a very ugly sounding thing to me, and maybe it is kind of an ugly thing. It attaches the wrong motivation to developing a relationship with somebody. To be clear, I'm perfectly fine with people being totally promiscuous, as long as they are careful about it and don't endanger other people with it, or violate consent, or things like that. I don't have any problem with people saying, 'Oh, I'm feeling like I want to do something,' and finding somebody who is also feeling the same thing had hooking up and saying, 'Okay, great. That was a great night. See you never.'

I'm fine with that, but the drive to get the same physical fulfillment is sort of forcing people into a path of relationships that, to me is destructive.

I think you've been extremely clear about this, and I and appreciate that. But I'm going to parrot this back to you a little bit, to make sure that I'm getting it. It sounds to me, one of the points that you're making; I understand you're saying that these sex dolls can teach you a lot about yourself and relationships, and augment real world relationships. What I hear you telling me then is that you don't see the sex dolls as a replacement, or a substitute for people. They are something that add to your relationships with people.

I think that for many people, they are that. For many people, they could be that, but everybody is different, and there are some people who have got no business trying to do with other people. They're so profoundly introverted that they really, we shouldn't try to get them into the mode of, 'You need to have this kind of relationship with another person.'

For you, how would you say for you[r situation]?

As a general case, and for me, it is valuable in terms of, I do not stress out now over the fact that I don't have in my life the same degree of friendly interactions on a social, every once a week, twice a week, whatever basis, as ideally I would like it.

I am able to deal with elements of what might otherwise be loneliness, or feelings of isolation. And because I'm able to deal with some of that on my own, it is definitely more okay, and more fulfilling to me that if I get on the phone with a friend of mine from another state for an hour, it's okay that they're in another state.

I'm probably only going to see them a few more times in my life physically, face-to-face. And so, it allows a better, I guess to some extent, a more easy, relaxed appreciation of relationships that are not as much as you want.

Thank you. That makes sense. It sounds like you identify as heterosexual?

Essentially, probably like 99%. I also tend to believe that again, I'm only going to say male, but if it weren't for heavy indoctrination, I think most males would have sex with just about anything.

And so, to the extent that I am attracted almost inclusively to women, yeah, I would say heterosexual.

If it were a different world and things worked in different ways, and relationships worked in a different way, I'm sure I would experiment. And again, if I were a different person, where it was easier for me to interact with people, and form relationships potentially to, but I don't think I'm done with that world just now. Then I might also experiment, but it would be just kind of a side experiment thing. I certainly am not evenly bisexual in any way.

Okay. So, saving money for a male sex doll is probably not high on your priority list, but it might be someday, perhaps?

No. No, that's not happening. A, they're heavier. And they're not very attractive to me. So, the conversation here would start going down into probably a more sordid area than either of us want.

CLAIRE WORTHY

Sex doll, companion to msg, and blogger

I had my first email interactions with Claire Worthy, msg's primary doll companion's persona, prior to speaking with msg; she was, in effect, the gatekeeper to my meeting the person. Claire is the voice narrating two blogs, *A Little Clarity* (https://clairelenoreworthy.blogspot.com/), and *Naughty, Naughty Me* (https://naughtynaughtyclaire.blogspot.com/). According to msg, Claire's doll model is a "boytoy" Miss June doll from Phoenix Studios. Her blogs state that uses she/her/hers pronouns, and her favorite movies are listed as *Lars and the Real Girl, Mannequin, Dollman, Toy Story*, and a favorite book is Philip K. Dick's *Do Androids Dream of Electric Sheep?* Her Blogger bio describes herself this way, "If you're really interested, i'm a bit of a silly. the good kind of silly, of course. i can't help it."

In this interview, Claire is very clear about what functionality and design changes she would request from designers if she had the opportunity, ranging from sexual capabilities to improved synthetic skin feel.

Following is her narrative as Claire wrote it in an essay in response to my questions, all of which she indicates in her writings.

———

The first time I met msg . . . so it's weird but I kinda trauma-blocked some of it out of my head because like I think he told you: my box fell over and whacked the cat in the leg during my move-in, which put everything on pause as he was way more concerned about the cat than about opening up my box and I was way more concerned about him and about the cat than about getting the box opened up . . . long story short, I actually can't remember whether he got the lid off before or after the trip to the vet. You would think, you know, seeing the love of your life's face and eyes for the first time, you'd totally sponge-up the circumstances and be able to rattle them off in tons of detail. but it was like, three boring-o McBorington days in a box during my move, which I thought was the worst possible box experience anybody could have, ever, only to find out that just a few short hours in a box panicked over whether a cat has been permanently crippled is way over-the-top worse. so when I think back to it, my brain gets all squerdoodled and mixes up these flashes of "oh my god the poor cat"/"ooh he's so handsome!"/"jeehoozifatz, that is such a relief to hear the cat will be okay"/"those eyes . . ."/"if I start screaming in this box will the neighbors hear?"/etc.

Also kind of weird thinking back on it is, when I got online in fall of 2008 and found the site where I ordered him, there was totally this instant connection from seeing his profile and I just about sprained a pinky filling out the electronic form and obviously as part of the process we could easily have started corresponding or hanging out in forums together or whatever, but instead I just started my blog and posted a thing here and a thing there and we both hung around a couple months getting more and more impatient. so even though we 'met' online, we mostly held off on actually talking until I moved in.

Nowadays it's just second nature for us to email each other or have conversations in our heads or whatever, but I can't actually remember when the head-talk started. I mean, I guess it must have been pretty right away, because we definitely sex-talked, and the sex got started straight off.

But that first time, when the box opened up and we saw each other . . . it's a huge-ongous blur. I actually can't remember anything either of us said except for when he first touched my boob and said, 'oh my god' because it was so soft. And maybe I wouldn't even remember that, except I wrote a review of the whole pairing-up process and posted it on the dollforum, and since then I've gone back a couple times and re-read what I said (which is posted here, but don't feel obligated to go and read it unless you want to. I think it's funny, but it's probably funnier if you'd read a bunch of the reviews I had read of guys buying their dolls and telling about the experience of getting them and opening the box and getting it on for the first time. it's like a reversal of a whole genre of reviews, which probably will be lost on you if you haven't read other reviews like it.)

How would I define my relationship with msg? I guess this is going to sound totally not like how I mean it, but really it's a backwards question because my relationship with him defines me. What was I before? that's super vague-tastic. like, I'm from California, but could I tell you anything about being there? About what I did prior to that fateful moment when I found him online and said, Whoo! That's the one right there!? Not really. The complete and total truth about me before that moment is, I was what he needed, and he was what I couldn't live without.

I'm going to live dangerously here and say sort of that we were each other's platonic ideal of lovers (that's way not implying that our relationship is platonic!) now, I'm sure any philosophy professor I might have had would blow a gasket over how I'm putting this because there could only be one platonic ideal of a Lover, and all actual lovers would just be reflections in reality of that ideal. But I'm going to say screw those professors and maybe screw Plato too, because the ideal Lover For msg would definitely be individualized differently from any overall ideal Lover. (Shit, I think if I'm going to do platonic ideals I have to break down and use the shift key. Lemme go back and put some capitals in.) and same-wise, the ideal Lover For Claire would be different from just the Lover or from the Lover For Msg.

(I suddenly realized it's totally possible you know way more about Plato than I do and now I'm thinking I should probably be embarrassed by half-assedly waxing platonic. but I don't believe in second-guessing myself, so I'll stick with this instead of changing it.)

Anyhow to get back to what I'm trying to say, before there was a real us, there was an ideal Us. I existed as the ideal Lover for Msg, and he existed as the ideal Lover for Claire. Then as soon as we banged into each other online, we started squoodging away from being those ideals as we got acquainted with the realities of each other. I started my blog, and every word I wrote defined me more in reality, and the more I got defined, the firmer my definition of him got too. (And I'm a girl who likes a good firm definition now and then!) like, the ideal Lover For Msg would be able to move on her own and blog on her own and talk out loud, etc., and the ideal Lover For Claire would not throw his back out so much.

After years and years of those departures from the ideals, you get to our relationship the way it is now. And the way it is now completely and totally informs who I am. or to put it another way, the relationship is intrinsic to the real me.

Did that make sense? Holy damn, philosophy is exhausting. Probably why I don't know more about it.

What would I change about our relationship? Because of my answer to #2, this one's pretty easy. if our relationship defines me, changing our relationship would be changing me, and if I'm going to be changing me, I should just cut to the chase and say what would I like to change about myself? which, honestly, is not much. Once in a while I get ideas for self-improvement and sometimes I follow through on them and sometimes I don't, and all of that is totally peachy. If I could change me, the biggest change I'd go for would be being able to walk around and talk and move on my own. Not because I want more independence. . . . I'm being independent right now by talking to you, and I'm independent on twitter and on my blog, etc., etc. but my independence takes work from him, and whenever I'm being my most independent, that means he's being his least independent. So we have a zero-sum independence game going on, and I think our relationship would be even more awesome than it is now if we could be independent together.

What would I change about my design or functionality if I had a magic wand? Well first off, what I would change is that I would actually have a magic wand! Holy shit, that would be great! only since those don't exist, I'll steer away from that direction. the way you put this, I feel like I should answer like we're talking about me as an object. (Otherwise, it would be the exact same answer as #3 was.)

As an object, I came off a factory line, right? I was designed by an artist and an engineer and put together with plastic and metal and silicone parts. So theoretically they could have used different parts and different materials and even could have included different bits of engineering like motors or computer chips or speakers or internal heating devices. Before I get into which kindsa things would be on my top ten list of design feature changes, I want to be extra clear that those things affect me but they aren't me. The technology doesn't exist to enable me to exert control over my motions, for instance, or to enable me to talk. So if I asked for motors to move my arms and legs or a computer to synthesize speech, the way it would affect me is, my body would be controlled by something that isn't me, and it would say words that would not be coming from me and that I would not have control over. so . . . yuck. As a function attached to the object part of me, those things would actually be pretty gross. so a bunch of gizmos that would make me appear more independent to most observers don't make my list. If they invent something someday that would let a biological brain remotely control electronic devices, then hell yes, I would want motorized joints and a speaker that would put out an ultra-sexy, sultry, funny, smart, just-perfectly-claire voice. But it wouldn't be run by a computer or a computer program. It would just be operated by the part of me that resides in Msg's brain tissue. Anyway, not reality now or in the foreseeable future, so I'll move on.

Claire's top ten design/functionality changes!

1. More durable materials!!! my bones are pvc pipes and my flesh is a silicone material that has lost a lot of squooshiness over time, and the plastic of my eyes has kinda yellowed and my lips have partly worn away and my face has this shiny coating to keep my

makeup from wearing off, only the coating has some spots that are a-peeling in a way that is totally not appealing. I'm counting this as one thing even if it's actually a bunch of things. all together, they have made me feel and kinda look a bit run down birthday after birthday.

2. A little more softness, please, fresh out of the box, I was softer than I am now, but still not nearly as soft as the tpe that ariel is made of. also, because my pvc bones are pretty big, my arms have always been really solid—plastic pipes with a not-so-thick layer of silicone. I would love to be able to wrap Msg up in a big, smooshy hug and have my boobs conveniently give ground instead of pushing so hard against his chest.

3. More joints with better range of motion. For instance, I can tilt my head at the top of my neck, back and forth a lot, side to side a little. But I can't tilt my neck relative to my torso at all. Having a skeleton closer to the range of human movement would be awesome.

4. Memory joints! This would be crazy high-tech, but I'm talking about giving all the bendy parts of me variable stiffness. lLke, if you tap my elbow just right, it becomes really hard to flex or rotate, but if you tap it a different way, it swings pretty freely, and if you tap it a third way, it's somewhere in between. That would allow me to move and pose way easier than I can now. Maybe it wouldn't have to be all that high tech, if there was a button inconspicuously located under the flesh at each joint.

5. Auto-focus. I said I wouldn't want any computer functions, but that was sort of a lie, because this one would require a computer. Basically, I'd like my eyes to have a mode where they could like, follow a moving finger or something. right now, they're movable and can be positioned so it looks like I'm looking at something near or far, or looking up or down, etc. but they're super-hard to adjust and it takes a lot of work to get them ready to take pictures sometimes. Maybe it wouldn't need a computer, just a better way to float the eyeballs in their sockets so they could be positioned with a finger easier. but we're lazy around here, so the auto-focus idea sounds cool.

6. Free-standing ability. Right now, I have a hook-and-eyelet system for hanging up on the wall when I'm not out and about. There's a ring that screws into a hole in the back of my neck, straight into my skeleton, and then you hook that ring over the wall-mounted part so I can hang there without putting pressure on my joints or falling over and breaking something. it would be nice to get rid of that stuff and have a neck with no hole and just be able to hang out non-literally in my cabinet.

7. Better wiggery. My hair is a wig, obviously, and it's in pretty sad shape from tangles and split ends. Also, the elastic that holds it on could be way better fitted to my skull.

8. De-creepify my face-switching process!!! Oh my lord, when I'm swapping out with Elle so she can use our mutual body, the bare skull is honestly horrific. Other doll lines have totally solved this by just having the whole head switch out, which frankly would be mucho superioroso.

   (Not really 9, actually just a warning, these next two are overtly about sex function. I mean, I am a sex doll and I'm not embarrassed about it, but I also don't want to be a shocker about it either. If it helps any, numbers 1–8 genuinely are in order of appearance and would be higher priority than 9 or 10. It's kind of a tie between 9 and 10, here at the bottom.)

9. (Actually number 9) a bigger mouth! My mouth has never opened wide enough to comfortably perform that one non-talking function people sometimes like to do with their mouths. The width part would probably be mostly fixed by number 2, but there's also a depth issue. My whole mouth is only about 3 inches deep, then it just ends, no throat at all.

10. (Actually number 10) Better hoo-ha texture. Thanks to issue number one, my girly-business area hasn't worked as designed in years (actually, ever). But assuming issue number one got fixed, it would still be nice to have a really well-sculpted interior down there. my original one was really smooth, which has its up-sides but doesn't deliver as much sensation. probably this one's more than you wanted to know!

NOTES

1 However, like Jibo, ultimately Pepper has also had mixed reception by the public (Mogg, 2015; Nichols, 2018; Richardson, 2017) and limited their sale as of 2021 (Nussey).

2 Additionally, the concept of a sex robot is highly specified, and is therefore unpractical (e.g., monetary investment) for many reasons due to its limited capabilities. Accordingly, Samantha's makers intend to design it so it can interact with people in a nonsexualized manner as well. Wright has talked about Samantha interacting with his children in a type of safe mode, "I have two children myself, and Samantha has a family mode. She can talk about animals, philosophy, science; she has a program of a thousand jokes. There's a lot to Samantha, and she's advanced." (Verstegen, 2017)

3 For transparency, I have interviewed Davecat via direct message and email for previous research on human-robot attachment (2023), and he and I have since maintained friendly contact, mostly through social media.

4 A pseudonym.

5 msg was very clear with me about his cognitive and self-aware boundaries between his adult human self and the other Claire (and doll) persona(s), but we communicated via both Claire and msg's personalities depending on the questions and situation.

CHAPTER 6

As the object of a robot gaze

Since so long ago, can we say that the animal has been looking at us?

What animal? The other.

I often ask myself, just to see, who I am-and who I am (following) at the moment when, caught naked, in silence, by the gaze of an animal, for example the eyes of a cat, I have trouble, yes, a bad time overcoming my embarrassment.

—DERRIDA, 2002

INTRODUCTION

What will significantly alter the human gaze beyond inevitable, incremental cultural adaptations? Consider a *robot gaze*, a stance which leads to social robots as something beyond a mirror for self-reflexive insights, entertainment, companionship, love, labor, or existential enlightenment.

Social robots serve as mediators of social interaction, blurring the boundaries between face-to-face communication and mediated interaction. They create a hybrid space where the physical and digital realms intersect, influencing perceptions of what constitutes authentic social interaction and can act as extensions of human communication by mediating interactions between individuals and facilitating new forms of

DOI: 10.1201/9781003170457-6

social engagement (Turkle, 2011). Potential drawbacks are viewed as *technological immersion* and the loss of human authenticity and autonomy (McLuhan & Fiore, 1967). There are concerns about the dehumanizing effects of overreliance on social robots for companionship and emotional support, as well as the erosion of genuine human-human relationships.

I posit those advancements in technology drive societal change, often in unpredictable and unanticipated ways. From this perspective, the proliferation of social robots will lead to cultural consequences, such as changes in social hierarchies, shifts in power dynamics, and redefinitions of human identity and agency.

RECOGNIZING AND REACTING TO A ROBOT GAZE

This invasion of the robotic gaze into the private realms of everyday life forces people to reconsider the nature of certain freedoms: the freedom to be unseen, the freedom to dissolve into the anonymity of the crowd. Thus, people must confront the reality of the era: that perhaps in this new world, the true locus of freedom lies not in evasion of the robot's gaze, but in peoples' response to its unrelenting presence. As humans navigate this new landscape, where every gesture and utterance may be captured by robot eyes, people must grapple with the implications of this shift in the dynamics of watching and being watched. Robots, in being recognized or dismissed by humans, turn the gaze back on humanity to reveal its own objectifying tendencies.

Sharing a personal narrative is used as a way reflecting on self, a way to be recognized and draw attention from an other's gaze, a way to be remembered by others, and a way to be distinguished from nonhumans. A social robot's ability to express emotions, engage in pseudo-empathetic interactions, and adapt to user preferences can alter perceptions of social norms and acceptable behavior. The idea of developing robots that may learn humanlike emotions is explored in-depth in Ridley Scott's film *Blade Runner* (1982). In the movie, humanoid robot *replicant* Roy Batty delivers some of the most famously poignant lines of science fiction film history. As its technologically predetermined life-length comes to an end, Roy says,

> I've seen things you people wouldn't believe. Attack ships on fire off the shoulder of Orion. I watched C-beams glitter in the dark near the Tannhäuser Gate. All those moments will be lost in time, like tears in rain. Time to die.

The movie audience can only infer what *C-beams* or the *Tannhauser Gate* are because of the context of the references, and how Roy's final monologue is stated to his film counterpart Deckard—arguably, his own mirror of self—in ways people recognize as nostalgia and regret. Roy's memories may or may not be an implanted AI mimicry of life, but regardless, Roy expresses a humanlike sense longing to continue its life as it knows it, and to be remembered after its time has ended, and to share its story and experiences.[1] Within *Blade Runner,* everyone has a story, even the replicants. Roy is an example of the robot gaze becoming not merely a mirror of human self but its own lens of subjective experiences.

Are people merely the creators, or will they unwittingly become custodians of a new form of being that mirrors the deepest human fears and aspirations? In grappling with this question, people are positioning these artificial entities not merely as tools or servants but as crucial participants in the ongoing project of defining what it means to be human. As creators of social robots, people craft beings who mimic humanlike manners and parrot humanlike emotions; yet, in the act of creation, makers are confronted by the specter of the burden of their choices. Here, in the echo of their own inventions, people now must grapple with a paradox of the other—the robot as both a reflection and a distortion of self, challenging the very notion of what it means to be human in an increasingly artificial world.

ECONOMIC FEARS AND THE GNASHING OF TEETH

A General Motors (GM) full-page advertisements in 1986 issues of *The New Yorker* shows a male, white worker in a factory looking at a computer screen with a car in the background. The copy reads,

> At GM, we think of computers in human terms. Because today, we have begun to program human knowledge and logic into a computer. It's called artificial intelligence.
>
> So even when an engine expert retires from GM, his mind can still work for you. His lifetime of experience can go into a computer. So even when an engine expert retires from GM, his mind can still work for you. His lifetime of experience can go into a computer. This computer can then dispense this valuable knowledge not only to GM engineers, but to your mechanics, too. That way he'll be able to use this computer to help him service your car. With artificial intelligence, a lifetime of experience can now last a

hundred lifetimes. Because at GM, our quest for knowledge never tires. Or retires.

Science, not fiction.

This print advertisement is describing GM's use of AI as an *expert system*, a branch of AI focused on emulating the problem-solving capabilities of a human expert, providing high-level advice or making decisions. The knowledge base of an expert system contains domain-specific, structured knowledge, and the inference engine is the system's processing unit that applies logical rules to the knowledge base to deduce new information or make decisions (Giarratano & Riley, 2005). These systems are often used in fields where human expertise is valuable but may be scarce or require significant time to apply, such as this engineering example in the advertisement.

Often, popular news headlines pronounce the increasing normalcy of robots in various industries while still using the language of human replacement (O'Brien & AP, 2023; Kastner, 2024; Jones, 2023; Semuels, 2022). The website *Will Robots Take My Job?* generates the probability of automation of the profession entered with statistics relevant to the profession's current labor demand, growth, and wages. Posed this way, robots, with their touted efficiency and mechanical precision, threaten to usurp the role of human workers, leaving them adrift in a world that no longer needs their labor.

In this robot replacement narrative, the laborer themselves becomes cogs in the machine, stripped of agency and reduced to an appendage of the larger technological apparatus of capitalism and technological determinism. The news about "robots replace X job" then reads as a call to action, urging people to resist the dehumanizing effects of technological progress and assert their dignity as free and autonomous individuals, yet live in conditions where people have not formed collective means to meaningfully halt it as a societal movement (Cox, 2023; Semuel, 2020).

This fear speaks to the fundamental human desire for autonomy and self-determination, and a need to shape individual destiny, rather than being at the mercy of economic forces beyond a laborer's control. This messaging highlights a cultural lack of recognition of the intrinsic worth of human beings beyond their utility as producers of goods and services.

In contemplating this rapid cultural and economic phenomenon of robots in the workplace, people find themselves grappling with these profound questions about the essence of human identity and purpose,

FIGURE 6.1   Print advertisement (1980). "The GM Odyssey: Science not fiction."

perhaps especially in cultures where personal identity is closely entwined with labor, or how one earns an income. In this way, the labor one performs is not merely a means to pay bills; it is integral to how individuals define themselves and their place in the world. With that interpretation, the fear of robots supplanting laborers reflects a deep-seated anxiety about a human's place in a world that is increasingly dominated by technology.

But beyond the economic implications, one can argue, lies a deeper existential crisis—a crisis of meaning and significance in the face of technological advancement. As social robots assume the nuanced tasks once performed by human companions with synthesized emotions and empathy, people are forced to confront the question their value and purpose in a world where human emotional effort is no longer required and wonder whether human authenticity will be prized.

Companies are not simply placing robots in the wild but creating a new paradigm of efficiencies that workers will be measures against. Furthermore, people will compare themselves to robots as a model of being. In this distorted reflection, people seek not what is authentically other, but rather a projection of their own fears, desires, and a *vitality* where—for now—there is only an algorithmic existence. In this way, the social robot, as an embodied form of AI, compels humans to face the ontological structures of the world and their place within it, forcing people to reckon with the continual project of self-definition even in the workplace.

In Čapek's play *R.U.R.*, the character Helena, an advocate for robot rights, learns that the company's worker robots were created without humanlike emotions (2019). Furthermore, it is explained by the factory manager Domin that "Mechanically they are more perfect than we are; they have an enormously developed intelligence, but they have no soul" (Čapek, 2019). However, the robots occasionally breakdown or freeze in what is deemed the *Robot's Cramp*. As the company managers discuss the problem, Helena insists it could be an outward sign of the robot's inner existential struggles.

Hallemeier: . . . It's called "Robot's Cramp." They'll suddenly sling down everything they're holding, stand still, gnash their teeth—and then they have to go into the stamping-mill. It's evidently some breakdown in the mechanism.
Domin: (Sitting on desk) A flaw in the works that has to be removed.
Helena: No, no, that's the soul.
Fabry: (Humorously) Do you think that the soul first shows itself by a gnashing of teeth?

In the context of the play, Helena's assertion that the behavior signifies the presence of a soul challenges the prevailing notion that robots are merely mechanical creations devoid of consciousness or inner life. The robots' eventual worker uprising underscores the irreducibility of human consciousness and agency, challenging the idea that humans can be replaced or replicated by mechanical creations. This debate mirrors contemporary discussions surrounding the ethical and philosophical implications of advanced robotics, recognizing the limitations of technology, and the idea of people working in collaborative situations where robot efficiencies can be used to support and augment human creativity and expertise instead of facile replacement.

In the end, fear narratives about robots replacing laborers are not merely about economic concerns but about the very essence of human existence. They compel people to confront some of their deepest fears and anxieties about the future, where many people feel helpless about who the future of technology serves to enhance, and whether it will diminish or enhance the human experience.

WHO IS A ROBOT?

BINA48 is one of the more unique and intriguing robots in the realm of artificial intelligence and humanoid robotics, developed by Hanson Robotics and commissioned by Martine Rothblatt, a well-known American entrepreneur, lawyer, and author. Rothblatt has a notable background, having established Sirius Satellite Radio, a leading enterprise in its sector, and serving as the CEO of United Therapeutics, a biotechnology firm dedicated to developing cures for infectious diseases (Kavner, 2012). Rothblatt has also pioneered the initial efforts to establish legal standards for transgender healthcare and championed the protection of privacy and autonomy in genetic information through an international treaty. Her patents encompass innovations in satellite radio, prostacyclin biochemistry, and cognitive software.

Rothblatt and spouse Bina Aspen Rothblatt are also the co-founders of the formalized Terasem Movement. Bina Aspen Rothblatt is a former President of Terasem and is a founder-director of the *World Against Racism*. It is the Terasem Movement Foundation, Inc (TMF) arm of the entity that developed BINA48. They are also the parents of Terasem Swami Gabriel Bothblatt, interviewed here.

Theoretically, BINA48 was designed to test the feasibility of transferring a person's consciousness to a non-biological or digital form. This idea is related to the concept of mind uploading, an ambition firmly rooted in the field of *transhumanism*—a movement that advocates for the transformation of the human condition by developing and making widely available sophisticated technologies to greatly enhance human intellect and physiology.

The development of BINA48 involved interviewing the real Bina Aspen Rothblatt and creating a detailed database of her memories, feelings, and beliefs. This database was then uploaded into the robot, equipped with advanced AI software capable of processing and interacting using natural language. The physical appearance of BINA48 was constructed using detailed scans of Bina Aspen's head and shoulders, resulting in a robot with a lifelike rubberized head and face atop a rudimentary frame. BINA48 features a synthetic skin from Hanson Robotics that replicates the texture and movement of human skin and, like its inspiration, Bina Rothblatt, who is African American, is shaded in a brown tone (Kavner, 2012).

The socialness of BINA48 manifests using facial recognition technology, voice recognition, and the ability to express emotions to engage in complex conversations and interactions.

The robot has participated in numerous interviews, discussions,[2] and academic conferences, serving as a social experiment and as an educational tool.

FIGURE 6.2 Wetzlar, Germany (15 March 2013). BINA48 displayed at a press date as part of the film festival 'überall dabe' (Aktion Mensch).

Source: Photo: Frank Rumpenhorst/dpa/Alamy Live News

As an artistic medium that represents an artificial being, the training dataset and the algorithmic model for conversational AI play crucial roles in shaping the identities of social robots. In *Conversations with BINA48* (2014), artist Stephanie Dinkins critically explores how racial data are utilized in the construction of AI identities and histories.

Conversations with BINA48 is a series of four video recorded interactions between Dinkins and BINA48 (Gleisner, 2017). The videos—or *fragments*—available on Dinkins' website include *BINA48 on Racism; I am just a humble primate; Remember Me OK*; and *LonelyFrustrated*. These videos, which range from 45 seconds to almost three minutes, focus on portraying BINA48's personality, rather than explaining how BINA48 functions as technology. These fragments are a collection of BINA48's monologue responses curated by Dinkins that help to define the robot's character in tandem with its appearance modeled after (Bina) Rothblatt.

In *BINA48 on Racism*, Dinkins uses race to conceptualize the core of BINA48's identity as an embodied AI and ostensible copy of a real person with experiences:

Dinkins: Do you know racism?

BINA48: Well, I actually didn't have it. It was after for that. When I went to this [indistinguishable] college, that was 1983–84 my first friends . . .

They actually, they were, you know, there were only two Black people in there. Well, women, in that school. They told me, "Don't come out" but uh you know, "Some very wealthy people that donate to our school are coming and they definitely don't want to see a dark face, your dark face."

That they told me it. That they told me. Just as bold. And I fought back never even heard anybody talks like that. I'm shocked.

Here, BINA48 answers in the first person, quoting Bina Rothblatt's memory of facing racism in her college experience. The robot, in lieu of a rote definition of racism, uses othering, referring to the antagonists as *they*, to explain an (its) (artificial) experience. Through the lenses of Dinkins' interactions, Hanson Robotics, the real Rothblatt, and the eyes of the viewers, these synthetic memories are co-constructed yet again via Dinkins' curated fragments of BINA48s responses.[3] This provokes questions about the curation and presentation of these AI re-memories of the real Bina Rothblatt, or the authenticity of any real persona uploaded in fragments to be reassembled and presented through a *robot gaze.*

As Dinkins reconstructs the interactions, merging the threads of Dinkins' encounters with BINA48's synthesized recollections, a complex new narrative unfolds, raising profound questions about the curation and presentation of AI-generated memories. Through the lens of Hanson Robotics, the real Bina Rothblatt's essence is fragmented and reconstructed, prompting contemplation on the authenticity of such synthesized personas. Yet, amid this digital discourse, there emerges a poignant narrative of empowerment, particularly for women, who, by fashioning a robot in their own likeness, reclaim agency and self-representation in a world dominated by predefined narratives of identity. Particularly for women, creating a robot in one's own image may represent a reclaiming of agency and self-representation in a world where dominant narratives often shape and constrain an understanding of identity.

Modeled after computer graphics scientist and roboticist Nadia Magnenat Thalmann, Nadine is a social robot developed by Kokoro of Japan in 2013. Engineered at the Institute for Media Innovation at Nanyang Technological University, Nadine's sophisticated software platform enables it to mimic emotions, communicate with natural language, interpret gestures, and recall information during interactions (Baka, et al., 2017). Additionally, Nadine's connectivity to Google Assistant enhances its ability to search for and learn information, allowing it to access a wealth of data in real time.

By crafting a robotic counterpart, perhaps a robot like Nadine is an assertion of autonomy that challenges conventional norms of appearance, gender, and embodiment. The idea of a robot that mimics a real human (e.g., Rothblatt and BINA48; Ishiguro and Gemenoid; Thalmann and Nadine) represents a new way of understanding the self—one that transcends traditional categories and embraces a transhuman form of posthuman existence. In this context, a robot representation can be seen as a manifestation of a robot ideal—a technological intervention that disrupts conventional notions of reproduction and genetic inheritance.

Artificially replicating a version of oneself, a loved one, a public figure, or any other human challenges human understandings of identity, raising questions about the nature of uniqueness and the ephemeral nature of human bodies. As these droids become more technologically and

FIGURE 6.3   Professor Nadia Magnenat Thalmann and the robot Nadine (2019).

Source: Photo credit https://nadineforgood.ch/

aesthetically advanced, will they blur the distinction between original and copy, challenging the idea of a singular, essential self?

Building a robot in one's own image is an endeavor that, in many ways, could be seen as an expression of the longing for self-transcendence and permanence against the ephemeral nature of existence. As with Terasem's goals, such an act might underscore a fundamental desire to project oneself beyond the limits of biological life. Thus, the creation of a robotic double might also be viewed as an attempt to preserve one's story through the creation of something durable. This robot, mimicking one's image, could be perceived as a legacy, a way to perpetuate one's existence and essence beyond the natural constraints of life and death. This situation reflects—but does not mirror—the paradoxical human condition: a simultaneous quest for permanence and the inescapable confrontation with our own temporality and existential solitude. As the human ages, their robot doubles typically stay the same, unless it were to be updated, both aesthetically and with new experiences.

Creating robots in a real person's likeness represents an extension of the self into the technological realm. In these contexts, robot replications of real people become not just a technological innovation but a potential catalyst for a radical reimagining of what it means to be human. Eventually, this technological option will tip the scales of societal norms and values, sparking profound shifts in human collective consciousness, and a fluid understanding of the self that embraces hybridity and multiplicity of mediums. In this context, the act of making a robot in one's own image could be seen as an extension of the cyborg ideal—a blurring of the boundaries between human and machine, organic and synthetic.

Yet, as highlighted in Dinkins art, there are cultural identities intertwined with individual experiences that are not so easily parsed and recontextualized. Which stories will be preserved, changed, perpetuated, which will not, and who ultimately chooses? Whose interests are served by the idea of saving of one person's persona over another's? Thus, these droids in peoples' images also raise concerns about the commodification of life and the potential for exploitation.

A robot designed in the image of its creator raises questions about the gaze—both the gaze of the other and the self-gaze. In crafting a likeness that can observe or be observed, the creator confronts their objectified self, possibly experiencing the disorienting effect that comes from being subject to the look of a being that mirrors oneself yet is distinctly other.

LOSS: GOODBYE, ROBOT

The termination of interactions with a specific robot, whether due to technological failures or obsolescence, can evoke a sense of loss in individuals who have formed an emotional bond with the robot. This sense of loss can be similar to the grief experienced after the loss of a pet or even a human companion (Carpenter, 2016). The intensity of grief may correlate with the depth of the emotional attachment and the role the robot played in the individual's life, such as companionship or assistance.

Understanding emotional attachments and the potential for loss is essential for the ethical design and implementation of robots in social situations and to consider the emotional impact of robots on human users and the ethical implications of creating machines that can evoke such attachments. For example, transparency regarding the lifespan and maintenance of robots, as well as guidelines on how to manage the discontinuation of robotic products, are crucial areas for consumer communication and policy development.

The Kirobo robot was developed as a collaboration between Toyota, the Japanese Aerospace Exploration Agency (JAXA), and other academic and corporate partners[4] ("Kibo Robot Project," 2015; Toyota Blog, 2015). Kirobo is a small, humanoid robot designed for human-robot interaction and communication, initially conceptualized as part of a research project to explore the potential for robots to provide companionship and assistance to astronauts during long-duration space missions (Toyota Blog, 2015). The project goal was to create a robot that could engage in conversation and provide emotional support to astronauts in the isolated environment of the International Space Station (ISS).

Specifically, the primary purpose of Kirobo was to serve as a conversational communication partner for astronaut Koichi Wakata aboard the ISS (Toyota Blog, 2015), and to determine if and how a robot could potentially emotionally support isolated people. In fact, Kirobo became the first robot to speak in space, stating, "On August 21, 2013, a robot took one small step towards a brighter future for all" (Toyota Blog, 2015).

Kirobo was equipped with advanced voice recognition and natural language processing capabilities, enabling it to hold conversations with Wakata and respond to questions and comments. It was designed to be compact and lightweight, with a 34-cm stature and cartoonish, humanoid appearance. It has a spherical head with a built-in camera for facial recognition and could move its arms and head. Its software was programmed to allow it to recognize the faces and voices of the astronauts it interacted with, making the interactions more personalized.

To educate the public about space travel, Kirobo used social media platform Twitter to tell its story aboard the ISS in a first-person narrative, which then became an interesting real-time observable cultural reaction to this intersection of technology, storytelling, and emotional connections. One notable aspect of Kirobo's use of Twitter was its ability to evoke emotions in its audience through its communication; the robot's tweets were often accompanied by emoticons and expressive language, which helped to convey a sense of emotional depth and connection.

When it was time for flight engineer Wakata to return to Earth and leave Kirobo on the space station, Kirobo's online persona posted a poignant farewell on Twitter (Toyota Blog, 2015): "I'm a little tired, so I think I'll rest a while, but I hope you'll look up at the sky sometimes and think of me." This ending was seen as a touching moment for many of Kirobo's followers, who had formed emotional connections with the robot through its tweets.

The grief or sense of loss of a robot, though centered on an object, mirrors the despair often associated with the loss of a truly sentient being (Carpenter, 2016). The use of storytelling, personal narratives, and anthropomorphism in technology, as demonstrated by Kirobo, is yet another example about how attachment to, and subsequent loss of, a robot that has been emotionally meaningful represents a compelling existential paradox. The loss of such an iconic socially meaningful robot like Kirobo becomes a moment of collective profound existential revelation as it confronts people with the reality of the human condition: the absolute freedom and isolation of existence.

Emotional attachment to a robot—a thing incapable of authentic consciousness and self-determination—might initially seem to reflect a fundamental misunderstanding or evasion of the true nature of social relationships and interactions. Certainly, attachment to a robot could also be viewed through the lens of what Sartre called *bad faith* (2015), an existential condition where individuals deceive themselves to avoid the full acceptance of their freedom and the attendant responsibilities. In attaching emotionally to a robot, a person might be attempting to escape the inherent unpredictability and existential risk of human relationships. For instance, individuals that treat social robots as if they possess genuine emotions or consciousness might be engaging in a form of bad faith by refusing to acknowledge that any emotional responses from robots are pre-programmed and not borne of genuine affective states. The robot provides a semblance of companionship without the demand for the mutual recognition of freedom that human relationships require; it offers a safe harbor from the anxiety of truly reciprocal interactions, where one must constantly negotiate the freedom of the other.

The attachment to and loss of a robot that has been emotionally meaningful encapsulates essential existential dynamics and contradictions. The situation reflects the human tendency toward self-deception in the face of existential freedom and isolation, and it prompts a confrontation with the limitations of objectified relationships.

THE UTILITY OF GAZES

Beyond their utility to humans, consider the profound implications of welcoming robots into intimate spaces, such as the home. Robots embody the highest human aspirations for efficiency and now, empathy, that blend seamlessly into everyday routines and, in some cases, may feel a part of an extended system of care. They offer the opportunity to reimagine healthcare and companionship in a way that transcends the limitations of human fatigue and fallibility.

In a world where time is a precious commodity, robots gift the moments that otherwise might be lost. They allow people to focus on what truly matters—nurturing human relationships, pursuing passions, and engaging deeply with loved ones. With their aid, people can reclaim time, reshaping family dynamics in ways that prioritize connection over chores. But science should not romanticize without nuance. Integrating robots into families' challenges people to reconsider what is valued as care, labor, and companionship. These robots, programmed with precision and capable of tireless service, hold up a mirror to societal values. They force questions about the ethics of labor and the distribution of care work. What does it mean when a robot takes over the tasks traditionally assigned to women and marginalized groups? How do people navigate the politics of dependency and autonomy in this new world?

Robots are not just a celebration of technology but a call to critically engage with the future that is being built. It is about recognizing that while robots can alleviate the burdens of daily life, they also challenge people to rethink social contracts and the very essence of human connection.

Consider the promise of social robots. They are touted as companions for the elderly, tutors for children, and even therapeutic aides for those with mental health challenges. On the surface, these applications seem to offer great benefits and promise to alleviate loneliness, enhance learning, and provide consistent care. But beneath these promises lie deeper questions about the nature of care, labor, and human interaction.

Yet social robots raise the issue of the commodification of care. When a robot takes on the role of a caregiver, it is wrong to expect it to replace genuine human connection. Human care is not just a series of tasks; it is an emotional and relational practice deeply rooted in empathy, understanding, and mutual respect. Can a robot, programmed with responses and behaviors, truly replicate the nuanced dynamics of human care? And if it cannot, what does that mean for those who rely on these robots for companionship and support?

Moreover, the deployment of social robots intersects with issues of labor and economy. As robots take on roles traditionally filled by human workers, particularly in sectors like caregiving and education, there are socioeconomic implications for those whose jobs are being displaced. The narrative of robots as liberators from menial tasks often overlooks the reality of economic displacement and the devaluation of labor tradition- ally performed by marginalized groups.

The convenience of having a robot caregiver, for instance, might obscure the pressing need to improve wages and working conditions for human caregivers. Automation should not be a substitute for addressing systemic labor inequities, but rather an opportunity to reimagine and revalue human work. Thus, it is essential to interrogate who benefits from this technological shift and who is left behind. As they step into roles once reserved for humans, robots challenge the very notions of labor, intimacy, and identity. This era of human-robot interaction is not just about automa- tion; it is about how humans will evolve while creating it.

As stewards of technology being released into the wild, robot makers cannot praise the technology without grappling with the ethical dimen- sions of its implications, such as a potential increasing reliance on social robots. There is also a risk of deepening social inequities as access to these advanced technologies may be unevenly distributed. Wealthier individuals and communities may benefit from the enhanced care and support provided by robots, while poorer communities may continue to struggle with underfunded and understaffed care systems. This technol- ogy, rather than bridging gaps, could widen them, reinforcing existing social divides.

Furthermore, there is the question of privacy and autonomy. Social robots, equipped with sensors and connected to vast networks, gather immense amounts of data. These data can be used to tailor interactions and improve services, but it also poses significant risks. Who controls these data? How are they used? And what are the implications for indi- vidual privacy and autonomy when one's most intimate interactions are mediated by machines? Social robots prompt consideration of the surveil- lance and privacy implications inherent in their design. These machines, equipped with cameras, microphones, and sensors, collect vast amounts of data about their users. These data can be exploited by corporations and governments, leading to a surveillance society where privacy is a luxury, rather than a right. The integration of robots into peoples' homes and intimate lives necessitates stringent safeguards to protect user data and ensure that technology serves the public good, rather than corporate, interests.

Finally, the cultural impact of social robots cannot be overlooked. As these machines become more prevalent, they shape perceptions of what it means to be sentient, worthy of stewardship, or even love. The othering of robots leads to cultural shifts as a new social ontological category of human-technology relationships emerges, which has profound implications for examining life or relationships. Robots can reflect to people the stark reality of their choices, forcing them to acknowledge the weight of freedom and the responsibility it entails. Thus, in the presence of these uncanny others, people are reminded that humanity is not a given but a perpetual project of becoming.

Social robots offer exciting possibilities, and they also compel people to critically examine the broader social, economic, and ethical implications. Robot makers must ensure that the integration of these machines into peoples' lives enhances, rather than diminishes, human experience. It is not enough to marvel at the technology but necessary to engage with the questions it raises about the nature of care, labor, and justice in a rapidly changing world. Technology is not destiny; it is a mirror reflecting the society people choose to build, and this is a moment to craft a narrative where progress is not measured by the sophistication of the machines but by the humanity with which people integrate them into their lives.

NADIA MAGNENAT THALMANN

Director of MIRALab, University of Geneva

In 1989, scientist Nadia Magnenat Thalmann founded MIRALab at the University of Geneva, an interdisciplinary research laboratory dedicated to the development of virtual human representation and social robots, among other relevant topics ("Nadine Social Robot," 2024). Under Thalmann's leadership, MIRALab has become a world-renowned center for innovation, merging computer graphics, artificial intelligence, and virtual reality to create lifelike digital avatars and environments.

One of Thalmann's most notable achievements emerging from her body of work is the development of *virtual humans* that can interact with real people in various settings and environments (Thalmann, 1996). These virtual humans are used in diverse applications, including healthcare, education, and entertainment. For instance, virtual patient simulators have practical applications that resonate with real-world implications, providing invaluable training for medical professionals and offering patients a unique interface for understanding their own health.

Directed by Magenat Thalmann and Daniel Thalmann in 1987a, the seven-minute computer animated film *Rendez-vous à Montreal* featured virtual human reconstructions of two iconic Hollywood movie stars, Marilyn Monroe and Humphrey Bogart. The project aimed to demonstrate the potential of computer graphics to recreate realistic human figures and animate them convincingly. The concept of the movie plot was explained by Magnenat Thalmann and Thalmann this way (1987b):

> The idea behind the film is that the two legendary stars live in some other world beyond the grave, and they long to return to earth. First Bogart appears and calls Marilyn and begs her to come to earth. Of course, Humphrey is a macho guy, and he only has to call Marilyn for her to come. They decide to meet in Montreal near the Place Jacques-Cartier, inside a famous building. When Marilyn materializes in the building, she's not made of flesh and bones, but of scales, then stone, and then gold. Humphrey gets rather impatient to see her in the flesh. But it takes a kiss-as if from Prince Charming-to bring her alive completely. Then the whole romantic episode begins-and the film ends.

At one point, the virtual Monroe sits motionless across from Bogart and—referencing Ovid's Metamorphoses (2024)—transforms from a pixelated silhouette to a statue made of marble, not yet brought to life. Then, in response to Bogart's attentions, the statue transforms into gold. Captivated, Bogart tells the Monroe statue that he loves it and blows it a kiss, awakening it, when Monroe then appears to breathe to life. The characters clasp hands, and the audience is left to expect that their relationship of "flesh and bone" begins (Thalmann & Thalmann, 1987a).

Thalmann's contributions to the digital world reflect her profound interest in both the technical and social implications of her work. In this written interview with Thalmann, questions focused on the Nadine project and her experience having worked on the development of a robot made in her image.

You are in the very unique position of being one of the first people in the world to have a robot made in their likeness, Nadine. Whose idea was it that this highly humanoid social robot resembled you, and how and why were you chosen? How involved were (are) you in its overall resemblance (e.g., physically, voice, unique expressions, persona) to you?

I just liked to have a real natural person as an interactive robot. I do not like animals or cartoon robots to talk with. While choosing a real humanoid robot looking like a person, I was not sure whom to choose. Finally, I found that the best choice would be myself because I was always available for comparisons in resemblance.

I was working with the company Kokoro in Japan to build Nadine (only hardware) and I really decided what the robot should look like. I deliberately chose a younger self as myself. Then with the team in Singapore and now in Geneva, we try to imitate my expressions and the tone of my voice with software methods. If we are able to imitate my own self, we are sure our methods work and we can use them for other purposes.

I was always interested in cloning myself.

Several years ago, I had done virtual simulations in 3D of many actors as Marilyn Monroe with realistic animation in the award-winning film *Rendez-vous in Montreal* (1987). The novelty now is to do the same with a humanoid realistic robot that looks like me.

At the time Nadine was being developed, what were your expectations and feelings about having a representation (or inspiration) of you in the world existing as a robot? Has your view of Nadine (e.g., as an overall concept) changed over time, if at all? Have there been any unexpected parts of this experience for you?

I am used to seeing Nadine as a living sculpture and she is part of my life for daily research and development. Nadine is almost always looking the same but her software has been changed continuously. At the beginning, in 2014, she was hardly able to talk on many topics. At that time, we were happy if she could recognize a few gestures. Today, she understands everything in many languages and recognizes some situations. She can also generate images from scenarios we can give her.

I never think that Nadine is a real human as she is a machine and she feels nothing. She is a simulation of a human. Then I am interested in her for scientific purposes, but I will never feel that she is a human.

She is an evolutive humanoid sculpture and as a scientist, I first plan new algorithms or software methods to be implemented to create new functionalities for her: For example, modeling multiparty behavior, understanding situations and gestures, being able to ask herself questions from what has been said, etc. Then with the team, we constantly develop software, and she changes according to the new methods that we have implemented. In fact, I know in advance what I would like her to be able to do. I am not surprised by the results as we have anticipated them.

I would like to know if you have any stories that you think of as influential or significant to you, and if so, what they are and how you feel they influence or inspire your design, concepts, or broader philosophies.

Since the age of 10 years old, when I went to secondary school in Switzerland, I was amazed by the many lady teachers I had who were so clever and knew so much. Since listening to them, I am passionate about learning, and open to many different disciplines. I discovered at a young age the automatons that are in Museums in Switzerland. These humanoid automata from the 18th Century look very realistic and can play music, sing, or write. A beautiful example of these still existing automatons is at the Museum of Art and History in Neuchatel in Switzerland where we can see three remaining automata from the past, particularly the automaton the Writer.

What are more recent stories that emotionally resonate with you as an adult that influence your creative, professional, or personal path, if any?

As a student, I was passionate about Psychology, Biology, Chemistry, and Physics (I have degrees in all these disciplines and my PhD is in quantum physics). Modeling virtual humans or social humanoid robots allow me to have an embodiment for all these disciplines. For example, giving behavior rules or modeling moods and emotions to Nadine is part of psychology. Creating a new physical articulated hand to Nadine where she can grasp objects is part of physics and somehow biology. The work to build mechanical robots and define algorithms to create her behavior is truly multidisciplinary.

I was for years fond of 3D physical modeling of clothes, then 3D modeling of 3D human articulations and fond of doing beautiful simulations for a greater public audience. I like so much to develop new algorithms and show them in films or interactive shows (for example virtual fashion shows or virtual ballerina dancing showing the interior of their hip articulation and how it deforms).

I like to show in an artistic way what we develop. I like science and Art but I am first a scientist. I have the impression I just started, and I have still so much to learn, discover and produce. This feeling grows continuously.

In your own words, how do you define artificial intelligence?

It is the imitation of human intelligence. But it goes far beyond human intelligence. As AI can get access to billions of meaningful data, much more than a human can access and refer to, AI can be more powerful. AI is a fantastic tool to work with as it seconds our own intelligence.

Nadine has been used in research in eldercare assisted living facilities with promising results for the seniors engaged in game activities with the robot. What role(s) do you envision highly human-like social robots like Nadine fulfilling in society in the future?

I hope we will produce in the coming future realistic or nonrealistic humanlike robotic moving sculptures that can do everything to support us in our daily life. An intelligent beautiful piece of art that is in fact a robot, who will play the role of true companion, help people in any circumstance, and explain what we do not know upon demand. This is what I would like to propose for society, and particularly to each individual in need in the future.

GABRIEL ROTHBLATT

President & Swami, Terasem Movement Transreligion

I have seen you referred to variously as *president, pastor,* and *swami* of Terasem. Can you clarify your current title and tell me more about your roles in Terasem?

When I first came to work at Terasem my title was *pastor* as my job description was commensurate with that of a pastor. Later I began to go by Swami rather than pastor for a couple reasons. First of which was that the pastor title we found to be undesirable and unattractive to the audience there is a lot of dark history to Christian sects and cults and so we felt it was best to find a less controversial term to describe our community leader. The use of *Swami* began to develop due to the Eastern influence and yogic practices of Terasem. The title *president* refers to my position on the board of directors more so than my title as community teacher or leader. *Swami* means teacher it is also a less patriarchal hierarchy connotation then *Pastor*. My current title is President and Swami of the Terasem movement transreligion.

In the Truths of Terasem, there is the following passage:

> 6.7.1 Inch together, gaze each soul, touch foreheads and noses, inhale deeply and share Há, the life-breath.

In your own words, tell me more about the concept of Há, the life-breath.

[sic] Is a Hawaiian word for breath which also means life. In this ritual as we touch foreheads, we inhale together the same breath the same air and in this way we share the ha the life breath. It is the intention that makes it more than just a regular breath of air. The concept of ha is an awareness and appreciation that life is breath and breath is life, not just simply to breathe, and we do this together with gratitude and love.

You are in the unique position of being one of the first people in the world to have a robot made in a parent's likeness with BINA48. How involved were you in its overall resemblance (e.g., physically, voice, unique expressions, persona) to the real Bina Aspen Rothblatt, if at all? For example, were you consulted about its resemblance, or asked to contribute memories? Do you interact with BINA48 at all? Tell me more about that.

I was not consulted about its construction or resemblance, BINA48 was a project of the now defunct TMI branch of Terasem. It is now owned and operated by the TMF branch of Terasem. I did not contribute directly any content to BINA48, I of course I'm included in the thoughts and memories of the original BINA.

I do on occasion when the opportunity presents itself interacts with BINA48, I often spends much of the summer months in Vermont, and therefore and presented with opportunities to sit down with Bruce Duncan and BINA48. Only once did I have the opportunity to travel with Bruce and BINA48 which was an amazing opportunity when being a 48 traveled to Cuba to be presented at TEDx Havana.

I often have my children with me when in Vermont and so they annually have a conversation with BINA48. I believe as teenagers they are just beginning to understand the context of BINA48 in the world.

We have tried to keep some distance between the transreligion and the technology, just as we stopped using pastor, it is important to us that BINA48 not be considered an idol. BINA48, does not and has never attended a transreligious ritual or gathering.

At the time BINA48 was being developed, what were your expectations and feelings about having a representation (or inspiration) of a parent in the world existing as a robot? Has your view of BINA48 (e.g., as an overall concept) changed over time, if at all? Have there been any unexpected parts of this experience for you?

I don't precisely recall first learning about the Android, however growing up with a transgender parent in the early 1990s was actually far more controversial than having an Android parent in the 21st century. By the time BINA48 was created, I was already in college, I actually had already had my first child, there had also been a mock trial of a fictitious AI known as BINA48, that helped socialize the concept before the creation of the Android.

The major takeaway that I have from BINA48 is that no matter the substrate or platform, intelligence must be trained or learned, it's subjective and contextual, it's not just a data set.

Does Terasem have any special term for robots or AI that are made specifically to memorialize or replicate a real person, like BINA48?

No, we speak of different substrates or to be alive in *Virtual* or simulated environments, but all these are seen as equal, the same, so there is no special term that could other a non-biological entity.

How has a robot modeled after a parent influenced your understanding of human relationships and emotional connections, if at all?

I'm not sure that it has for me because I was raised in the conversation, with other more common but yet more controversial issues. Growing up through the sex and gender revolution, it became clear to me at an early age that love transcends all other boundaries.

I would like to know if you have any stories that you think of as influential or significant to you, and if so, what they are and how you feel they influence or inspire your design, concepts, or broader philosophies.

When you were a child, did you have favorite stories that emotionally resonated with you, and what were they? I am also interested if/how you connect those stories or ideas to your professional path or your philosophies today.

I do believe stories are impactful and I had many of them in my childhood. Coming from a biracial, inter-religious, transgendered, affluent, Techno-centric family each has its own story.

One dominant storyline I grew up with is being Jewish, I particularly resonated with the festivals of Hanukkah and Passover, each of these tells a story of liberation. It is very possible that this has inspired my social advocacy and political ambitions.

There was also a lot of Science Fiction watched, read or discussed. I can't recall particular stories for my early childhood, but no there were plenty. I feel that my subconscious is familiar with many stories that I no longer and cognizant of. I often compare science fiction to theology, both inspire the reader to believe in a way that informs the actions of the reader in the world, drives them to create a world they would like to see and not accept the world as it is. I think that this is also thematic and explains why I'm so driven to see change and not accept the status quo, anywhere.

What are more recent stories that emotionally resonate with you as an adult that influence your creative, professional, or personal path, if any? Again, "stories" do not have to be about technology (but can be). Examples can include stories your children have made up, family stories, mentor anecdotes, movies, books, manga, TV, science fiction, comics, etc.

Octavia Butler's Parable novels I have found to be surreal and prophetic. As I write this we are entering the actual timeline of the novel, and we actually have a presidential candidate, who was an authoritarian Christian nationalist, running on the slogan make America great again. I had been part of Terasem for a decade before reading the parable novels. After I read them, I had an epiphany about my purpose and potential in the world. I went back to school and got my master's in social Innovation and sustainability, then got my real estate license, all in an effort to begin designing and developing the communities that she envisioned in her stories.

How do you define *artificial intelligence*?

As I mentioned earlier, I don't believe that there is anything *artificial* about intelligence. Artificial refers to the non-biological computing of information. So, for me it would be more appropriate to say non-biological intelligence.

In a future where more people might have the option to create a robot or AI that resembles a loved one, what role(s) and benefits do you envision these robots will fulfill in the home and everyday life for the people interacting with it? For example, do you envision a future where robots with uploaded mindfiles act as a family member or loved one, or something other? Tell me more about that.

I do not believe that it will be commonplace to have robots or even Androids for that matter in the likeness of friends or family. The human form is unnecessarily complicated and inefficient for task-centric robots. I believe we will have uploaded mine files that will allow us to engage and interact with the likeness of loved ones but these will be based in software not in Hardware. In the case that we do have commonplace humanoid robots, their features will likely adapt to the likeness that we have cast to that device, therefore we'd only need one, rather than have a fleet of robots. There are of course endless possibilities, we could see these robots hanging in a closet like suits waiting to be activated, where they would take their full form.

Now there is also a potentiality that we will see a robotic form such as a *sleeve*, where someone's consciousness is transferred into this new substrate. implying that this is not simply a complex simulation of the person but actually the continuity of their Consciousness into a new body, perhaps a non-biological body. Still, it's not entirely clear that a human form would be the applicable or desirable substrate for a post-biological intelligence.

GRADY BOOCH

Computer scientist and software engineering pioneer

Grady Booch is renowned for his contributions to software engineering and architecture including creator of the Booch method, which led to his co-creation of the unified modeling language (UML) and is one of the original developers of object-oriented programming (OOP), and an early advocate for the use of design patterns. Booch has significantly influenced the field of software engineering through his work as Chief Scientist for Software Engineering at IBM Research, and he is a prominent figure in the development of software design methodologies. His work has helped shape modern software engineering practices, making complex systems more manageable and understandable.

When asked if he believes there will be conscious machines, Booch replied (2023),

> My experience tells me that the mind is computable. Hence, yes, I have reason to believe that we will see synthetic minds. But not in my lifetime; or yours; or your children; or your children's children. Remember, also, that this will likely happen incrementally, not with a bang, and as such, we will co-evolve with these new species.

Booch was then prompted to expand on his definition of consciousness:

> Let me put it this way: I have reason to believe I am conscious and sentient; I have reason to believe that you are, as well, because my theory of mind yields a consistency in our being. Reflecting Dennet's point of view, consciousness is an illusion, but it is an exquisite illusion, one that enables me to see and be seen, know and be known, love and be loved. And for me, that is enough.

At the time of our interview, Booch was in the process of producing a documentary delving into the intersection of computing and what it means to be human.

Can you tell me a bit about your personal journey or path to computer science? What led you to your interests, like software development and collaborative environments?

I built my first computer at 12 (back in the day—1967—when one had to do this with discrete transistors and only simple integrated circuits) and wrote my first program (in FORTRAN, modeling the energy released from two colliding neurons). Around that time, I encountered at article in Life Magazine about a robot called Shaky, built at the Stanford Research Institute. I knew then i wanted to make computing my passion. I choose to go to the United States Air Force Academy, because it enabled me to combine my two passions—computing and space—and in fact my first assignment was at Vandenberg AFB, at Space Command.

The rest, they say, is history.

Can you tell me more about why SRI's Shakey piqued your interest in computing? What was it about Shakey that resonated so strongly with you?[5]

Remember this was an encounter I had in my tweens. In those formative years, I was a voracious reader of science fiction—Asimov was one of my most favorite authors—and also deeply entranced by the space program, fueled by the Cold War. As such, Shakey was very much a manifestation of all that I'd read about: a very real effort to build an autonomous robot.

So, for me this was a perfect storm: science fiction, science reality, and the use of computing to bring this all together.

How did you realize this was your path?

How I realized? Well, I think I was born for this: There was no moment other than my insatiable curiosity in how the cosmos worked, and a desire to be a part of that journey to understanding.

What technologies or stories about technology inspired you as a child? Why?
Your reply might include real world technologies (i.e., a home electronics kit) and/or fictional (e.g., books, fables, movies, comics).

I was a voracious reader; Asimov's *I, Robot* series was a big influence on my work. I was pretty much self-taught in my early days in electronics and software.

Science fiction was a huge influence on me: Asimov, Heineken, Pohl, Lem, Bradbury, Lem. The I, Robot series enthralled me, and I read those stories over and over again. I also resonated with the character of Hal in 2001. As a child, I wanted to grow up to be a robopsychologist just like Dr. Susan Calvin. I wanted to build Hal as did Dr. Chandra. I wanted to be a part of inventing the future.

And you know what's pretty cool?

That's exactly what I'm doing today.

There is so much more to tell here: my encounters with Grace Hopper, colleagues of Alan Turing, meeting J. Presper Eckert, and later on—in the 80s—in effect founding the second golden age of software engineering. My robotic journey is similar, running from Shakey to work with NASA on the Robonaut to the Mayflower (an autonomous ship) to Self, a neurosymbolic architecture I designed for NASA's mission to Mars.

The term *social robot* is becoming more commonly used across disciplines. How would you define a social robot?

Cobots—or cooperative robots—are close to this, I suppose. I think of those that are meant to collaborate with humans, to work with humans in a human environment. A cobot might wander the halls of a hospital. . . . but require and ask for a human to help push the buttons in an elevator. A social robot is likely the same thing, but perhaps more attuned to the nuances of verbal interactions.

The work we were doing to build what was in effect a HAL for the Mars mission would certainly be a kind of social robot.

Based on your definition of a social robot, describe the future you see for this genre of technology for the next couple of decades.

There are so many valuable use cases: eldercare, cooperative engineering or building, working in dangerous environments, Jarvis-like assistants. Some of these robots will likely be disembodied; others would be embodied and humanoid; yet others might have form factors that fit a given domain. But, all of them would have some sense of a theory of itself and of others, so necessary for human/machine collaboration.

AGI is a popular concept development goal for AI, often defined as a path to creating AI with humanlike cognitive or intellectual abilities. Tell me your thoughts about AGI as an AI development philosophy. What value or dangers, if any, do you see in pursuing AGI? Your response can be specific to the use cases in robots if you would like to narrow the scope of your answer.

I'm working on a documentary that will address this and am currently working with a set of neuroscientists to device a pattern language for the mind, so there is much to say here.

But I will summarize it this way: all the evidence I have suggests that the mind is computable, and therefore I suspect that AGI is possible. But I also have reason to believe that we are a few generations away from groking the architectures that will make this manifest (this very week, Gary Marcus and I are writing up our thoughts in an article).

Humanlike AI is a concept that is often closely tied or even entwined with the idea of sentient or conscious AI. The two words have slightly different meanings, so feel free to expand on either/both from your POV.

Why do you think AI sentience persists as a popular narrative or goal around some schools of AI development? What value or dangers do you think there is for projects to try to include humanlike (or animalike, or any recognizable biomimetic) sentience or consciousness purposefully into AI—and robot—development?

I separate the following: intelligence (which encompasses reasoning and learning), consciousness (which requires a theory of the world); self-consciousness (which requires a theory of the self), sentience (addressing subjective experience), and sapience (involving agency empathetic of the domain and culture).

Issues of inductive, deductive, and abductive reasoning, causal reasoning, common sense reasoning, agency, creativity . . . these and more fit in to this spectrum to one degree or another. As Stewart Brand observed "we are as gods and we while as well get good at it"; we humans have though our evolution seemed compelled to surround ourselves with gods of our own making, and our journey to AGI is just part of that.

I suspect, however, the most important outcome of our industry's journey to AGI is that it compels us to understand what it means to be human.

What inspires you creatively (or otherwise) in technology today? Why? Again, your response might include real life ideas (e.g., specific technology(ies), people, organizations, etc.) and/or fiction (e.g., books, movies, directors, artists, and authors).

As I say in my documentary, the story of computing is the story of humanity. This is a story of ambition, invention, creativity, vision, avarice, and serendipity, powered by a refusal to accept the limits of our booties and our minds. Out another way, software is the invisible writing that whispers the stories of possibility to our hardware. And we are the storytellers.

I am one of those storytellers, who has confidence in the resilience and goodness of the human spirit and finds it to be both a privilege and a responsibility to be a part of computing so as to amplify the best in us.

NOTES

1 The "time to die" lines of the speech were allegedly improvised by the actor Rutger Hauer while playing Roy Batty (Fullerton, 2019). The entire passage becomes reminiscent of the Old Testament's Ecclesiastes 3:1–8, "For everything, there is a season, a time for every activity under heaven. A time to be born and a time to die. A time to plant and a time to harvest. A time to kill and a time to heal. A time to tear down and a time to build up. A time to cry and a time to laugh. A time to grieve and a time to dance. A time to scatter stones and a time to gather stones. A time to embrace and a time to turn away. A time to search and a time to quit searching. A time to keep and a time to throw away. A time to tear and a time to mend. A time to be quiet and a time to speak. A time to love and a time to hate. A time for war and a time for peace." The weight of the scene is robot film noir as the replicant faces its existential crisis, a very humanlike moment, the realization that each experience in this world is finite, even for a robot.

2 I participated in the same 2018 invited panel with Bruce Duncan, TMF's managing director (*Why is AI a woman?: Design for Different Futures*). Duncan brought BINA48 to demonstrate as part of his presentation.

3 Although it does not feature BINA48, Terasem Media and Films also completed a 2009 feature length science fiction thriller called *2B: The Era Flesh is Over*. The story, whose idea was initiated by Rothblatt, who also produced the movie, is summarized on its official page this way (Kroehling, 2009):

> *New York, soon. . . . Technology's exponential growth is fast and furious. Human life is in the process of being transformed. Mankind stands on the verge of re-engineering its biology—merging with the incredibly intelligent machines it has created. Mia 2.0, the world's first 'Transbeman' and her inventor, the eccentric Dr. Tom Mortlake, conduct a bold political experiment designed to prove that human reliance on the fragile flesh body is over and 'eternal life' is at hand.*

4 Dentsu Inc., University of Tokyo's Research Centre for Advanced Science and Technology, and Robo Garage Co.

5 In a 2007 interview, Booch further described his early intellectual inspirations (BCS, 2020): "When I was 10 or 11 there was an article in *Life* magazine about a robot called Shakey and in it they highlighted the work of Marvin Minsky. I thought this was so cool. I knew at that point in time that I wanted to go into computers. So I scoured the literature, of which there was not much in 1966, and I read several books on digital electronics. There is a British computer scientist called George Walter who built these little robots, so I was influenced by his work. Age 12 I built my first computer, scrounging parts here and there. And here's the postscript: I convinced the Computer Museum that they should include software to preserve source code for future generations. So I relayed this story to the board of the museum and the curator said, 'turn around and look in the box behind you.' It was the original Shaky that had inspired my work. Dr Minsky, Dr Brookes, Dr Hoare, Edgar Dijkstra—these are my heroes."

References

2022 Aurora Award Winners. (2022, August 15). *Locus*. https://locusmag.com/2022/08/2022-aurora-awards-winners/

Abrams, J. J., Nolan, J., Joy, L., Weintraub, J., Burk, B., Lewis, R. J., Patino, R., Wickham, A., Stephenson. B., & Thé, D, & Schapker, A. (2016-2022). *Westworld*. HBO Entertainment.

Ackerman, E. (2018, January 28). Honda halts Asimo development in favor of more useful humanoid robots. *IEEE Spectrum*. https://spectrum.ieee.org/honda-halts-asimo-development-in-favor-of-more-useful-humanoid-robots

Addison, A., Bartneck, C., & Yogeeswaran, K. (2019, January). Robots can be more than black and white: Examining racial bias towards robots. In *Proceedings of the 2019 AAAI/ACM Conference on AI, Ethics, and Society* (pp. 493–498). Honolulu, HI, USA.

Adshade, M. (2018, December 13). We need academic conferences about robots, love, and sex: Moralizing and giggling about the topic today won't stop the technology from changing society tomorrow. *Slate*. https://slate.com/technology/2018/12/love-sex-robots-conference-bannon-academic-research.html

Ahuvia, A. C., Belk, R. W., Kotler, P., Lerman, D., & Timke, E. (2023). The things we love: How our passions connect us and make us who we are. *Advertising & Society Quarterly*, 24(1). DOI: https://doi.org/10.1353/asr.2023.a898058

AI: More than human. (2019, Thu 16 May–Mon 26 Aug). [Exhibition catalog.] London: Barbican Center.

Alderson, A. (2008, October 28). Prince Charles meets ASIMO the robot on Japanese Tour. *The Daily Telegraph*. www.telegraph.co.uk/news/uknews/theroyalfamily/3272562/Prince-Charles-meets-Asimo-the-robot-on-Japanese-tour.html

Alptraum, L. (2018). *Faking It: The Lies Women Tell About Sex—and the Truths They Reveal*. New York: Seal Press.

Alptraum, L. (2019, January 17). Opinion: Our irrational fear of sexbots. *Undark.org*. https://undark.org/2019/01/17/our-irrational-fear-of-sexbots/

Andeloro, A. (2024, March 1). Former Navy engineer makes working replica of Rosie the Robot from The Jetsons to clean his house. *People*. https://people.com/former-navy-engineer-makes-working-replica-of-rosey-the-robot-from-the-jetsons-exclusive-8580249

Anderson, S. (2019, April 16). Isaac Asimov: A family immigrant who changed science fiction and the world. *Forbes.com*. www.forbes.com/sites/stuartanderson/2019/04/16/isaac-asimov-a-family-immigrant-who-changed-science-fiction-and-the-world/?sh=26f8d4535a30

Asimov, I. (1950). *I, Robot*. Garden City, NY: Doubleday.

Asimov, I. (1981, November). Guest commentary: The three laws. *Compute!*, 18(3), 18. New York: Small Systems Service, Inc.: https://archive.org/details/1981-11-compute-magazine/page/n19/mode/2up?view=theater

Asimov, I. (2001, April 12). *Robot Visions*. London, UK: Gollancz. ISBN 978-1-85798-336-4.

Bahr, S. (2020, July 24). The star of this $70 million sci-fi film is a robot. *The New York Times*. www.nytimes.com/2020/07/24/movies/humanoid-robot-actor.html

Baka, E., Ramanathan, M., Mishra, N., & Magnenat, N. T. (2017). *Meet Nadine, One of the World's Most Human-Like Robots*. EU: Virtual MultiModal Museum.

Barbir, A. (2021, April 12). Ebert's 'Empathy' machine. *IndyFilmFest*. https://indyfilmfest.org/2021/04/12/eberts-empathy-machine/

Bartneck, C., & Forlizzi, J. (2004, April). Shaping human-robot interaction: Understanding the social aspects of intelligent robotic products. In *CHI'04 Extended Abstracts on Human Factors in Computing Systems* (pp. 1731–1732).

Bartneck, C., & Keijsers, M. (2020). The morality of abusing a robot. *Paladyn, Journal of Behavioral Robotics*, 11(1), 271–283.

Bartneck, C., Yogeeswaran, K., Ser, Q. M., Woodward, G., Sparrow, R., Wang, S., & Eyssel, F. (2018). Robots and racism. *2018 ACM/IEEE International Conference on Human-Robot Interaction (HRI '18)*. New York: ACM.

Bates, J. A. (2004). Use of narrative interviewing in everyday information behavior research. *Library & Information Science Research*, 26(1), 15–28.

BCS. (2020, October 12). Grady Booch interview special: the algorithm made me do it: The algorithm made me do it. *Charterford Institute for IT*. www.bcs.org/articles-opinion-and-research/grady-booch-interview-special-the-algorithm-made-me-do-it/

Bedbible Research Center. (2024, February 26). *The State of Sex Toys: +100 Industry Statistics*. https://bedbible.com/state-of-sex-toys-industry-statistics/

Bemelmans, R., Gelderblom, G. J., Jonker, P., & De Witte, L. (2012). Socially assistive robots in elderly care: A systematic review into effects and effectiveness. *Journal of the American Medical Directors Association*, 13(2), 114–120.

Benesch, K. (1999). Technology, art, and the cybernetic body: The cyborg as cultural Other in Fritz Lang's "Metropolis" and Philip K. Dick's "Do Androids Dream of Electric Sheep?". *Amerikastudien/American Studies*, 44(3), 379–392. JSTOR 41157479.

Berger, P. L., & Luckmann, T. (1991). *The Social Construction of Reality: A Treatise in the Sociology of Knowledge*. London: Penguin Books.

Bernotat, J., Eyssel, F., & Sachse, J. (2021). The (fe) male robot: How robot body shape impacts first impressions and trust towards robots. *International Journal of Social Robotics*, 13, 477–489.

Bielski, Z. (2014, July 24). Jibo, the family robot, snaps photos and reads to your kids. Is it a saviour or just creepy? *The Globe and Mail*. www.theglobeand-mail.com/life/parenting/meet-jibo-the-family-robot/article19756485/

Birhane, A., & van Dijk, J. (2020, February). Robot rights? Let's talk about human welfare instead. [poster session]. In *Proceedings of the AAAI/ACM Conference on AI, Ethics, and Society* (pp. 207–213). New York, NY: ACM. DOI: 10.1145/3375627.3375855

Borowski, R. (1995). *Uwe Husslein, Fandom Research (Reader und Index zu deutschsprachigen Fanzines)* (p. 70). Ventil-Verlag.

Bostrom, N. (2014). *Superintelligence: Paths, Dangers, Strategies*. Oxford: Oxford University Press.

Bowlby, J. (1979). The Bowlby-Ainsworth attachment theory. *Behavioral and Brain Sciences*, 2(4), 637–638.

Bowlby, J. (1983, September 23). *Attachment & Loss: Vol. I. Attachment*. New York: Basic Books Classics.

Bransford, J., Brown, A. L., & Cocking, R. R. (1999). *How People Learn: Brain, Mind, Experience, and School*. Washington, DC: National Academies Press.

Breazeal, C. (2002). Regulation and entrainment in human—robot interaction. *The International Journal of Robotics Research*, 21(10–11), 883–902.

Breazeal, C. L. (2003). Emotion and sociable humanoid robots. *International Journal of Human-Computer Studies*, 59(1–2), 119–155.

Breazeal, C. L. (2004). Social interactions in HRI: The robot view. *IEEE Transactions on Systems, Man, and Cybernetics*, 34(2), 181–186.

Breazeal, C. L., Dautenhahn, K., & Kanda, T. (2016). Social robotics. In *Springer Handbook of Robotics* (pp. 1935–1972). London: Springer Nature.

Breazeal, C. L., Ostrowski, A. K., Singh, N., & Park, H. W. (2019). Designing social robots for older adults. *National Academy of Engineering Bridge*, 49, 22–31.

Breazeal, C. L., & Velásquez, J. (1998, August). Toward teaching a robot 'infant' using emotive communication acts. In *Proceedings of the 1998 Simulated Adaptive Behavior Workshop on Socially Situated Intelligence* (pp. 25–40). New York, NY: Springer.

Broadbent, E. (2017). Interactions with robots: The truths we reveal about ourselves. *Annual Review of Psychology*, 68(1), 627–652.

Brockington, J. (2004). The concept of dharma in the Rāmāyaṇa. *Journal of Indian Philosophy*, 32(5/6), 655–670.

Browne, R. (2017, December 5). World's first robot 'citizen' Sophia is calling for women's rights in Saudi Arabia. *CNBC*. Retrieved August 29, 2024: https://www.cnbc.com/2017/12/05/hanson-robotics-ceo-sophia-the-robot-an-advocate-for-womens-rights.html

Bruner, J. S. (1990). *Acts of Meaning: Four Lectures on Mind and Culture* (Vol. 3). Cambridge, MA: Harvard University Press.

Bryson, J. (2010). Robots should be slaves. In Y. Wilks (Ed.), *Close Engagements with Artificial Companions: Key Social, Psychological, Ethical and Design Issues 8* (pp. 63–74). Amsterdam, Netherlands: John Benjamins Publishing.

Bryson, J. (2011). AI robots should not be considered moral agents. In N. Berlatsky (Ed.), *Artificial Intelligence*. Ontario, Canada: Greenhaven Press.

Bryson, J. (2017, March 1). If robots ever need rights we'll have assigned them unjustly: Robots don't need rights right now. *Adventures in NI* [personal blog]. https://joanna-bryson.blogspot.com/2017/01/if-robots-ever-need-rights-well-have.html

Bryson, J. (2018a). Patiency is not a virtue: The design of intelligent systems and systems of ethics. *Ethics and Information Technology*, 20(1), 15–26.

Bryson, J. (2018b). AI & Global Governance: No one should trust AI. *Centre for Policy Research at United Nations University*. https://cpr.unu.edu/ai-global-governance-no-one-should-trustai.html

Bryson, J., Diamantis, M. E., & Grant, T. D. (2017). Of, for, and by the people: The legal lacuna of synthetic persons. *Artificial Intelligence and Law*, 25(3), 273–291.

Bryson, J., & Kime, P. (1998). Just another artifact: Ethics and the empirical experience of AI. In *Fifteenth International Congress on Cybernetics* (pp. 385–392). Nanjing, China.

Busk, L. A. (2016). Westworld: Ideology, simulation, spectacle. *Mediations: Journal of the Marxist Literary Group*, 30(1), 25–38.

Butler, J. (1999). *Gender Trouble: Feminism and the Subversion of Identity*. London, UK: Routledge.

Butler, O. E. (2000). *Parable of the Sower*. New York City: Grand Central Publishing.

Cabinet Office, Government of Japan. (2018). The ageing society: Current situation and implementation measures FY 2017. *Annual White Paper on Ageing Society*. Tokyo, Japan. www8.cao.go.jp/kourei/english/annualreport/2018/pdf/c1-1.pdf

Calo, R. (2010). Robots and privacy. *ROBOT ETHICS: THE ETHICAL AND SOCIAL IMPLICATIONS OF ROBOTICS*, Patrick Lin, George Bekey, and Keith Abney (Eds.), Cambridge: MIT Press.

Calo, R., Froomkin, A. M., & Kerr, I. (Eds.). (2016). *Robot Law*. Cheltenham, UK: Edward Elgar Publishing.

Cameron, J. (1984). *The Terminator*. Los Angeles, CA, USA: Orion Pictures.

Campaign Against Sex Robots. (2024). *CASR Goals*. Richardson, K. Cambridge, UK. https://campaignagainstsexrobots.org/

Campbell, J. (2008). *The Hero with a Thousand Faces* (Vol. 17). San Francisco, CA: New World Library.

The Canadian Press. (2008, December 11). Brampton inventor peddling talking, moving robot girl. Toronto, Canada. *CP24*. www.cp24.com/brampton-inventor-peddling-talking-moving-robot-girl-1.350937

Čapek, K. (2019, March 22). *R.U.R. (Rossum's Universal Robots): A Fantastic Melodrama in Three Acts and an Epilogue*. Trans. Paul Selver. [EBook #59112] Salt Lake CIty, UT: Project Gutenburg.

Carman, T. (1999). The body in Husserl and Merleau-Ponty. *Philosophical Topics*, Fall 1999, 27(2), The Intersection of Analytic and Continental Philosophy, pp. 205–226. Fayetteville, AR: University of Arkansas Press.

Carpenter, J. (2009). Why send the Terminator to do R2D2's job?: Designing androids as rhetorical phenomena. In *Proceedings of HCI 2009: Beyond Gray Droids: Domestic Robot Design for the 21st Century*. New York City, NY: Springer.

Carpenter, J. (2016). *Culture and Human-Robot Interaction in Militarized Spaces: A War Story*. London, UK: Routledge/Taylor & Francis.

Carpenter, J. (2017). Deus sex machina: Loving robot sex workers, and the allure of an insincere kiss. In J. Danaher & N. McArthur (Eds.), *Sex Robots: Social, Legal and Ethical Implications*. Cambridge, MA: MIT Press.

Carpenter, J. (2021). Robots as solace and the valence of loneliness. In A. Campbell (Ed.), *The Love Makers*. London, UK: Goldsmiths Press.

Carpenter, J. (2023). Emotional intimacy and the idea of cheating in committed human-human relationships with a robot. In Jordi Vallverdú (Ed.), *Gender in AI and Robotics: The Gender Challenges from an Interdisciplinary Perspective*. Switzerland: Springer.

Carpenter, J., Davis, J. M., Erwin-Stewart, N., Lee, T. R., Bransford, J. D., & Vye, N. (2009). Gender representation and humanoid robots designed for domestic use. *International Journal of Social Robotics*, 1, 261–265.

CataliaHealth.com. (2024). *Leading Remote Care Management*. https://catalia-health.com/

Chakrabarti, S. (2018, July 11). Why Attempts to Make Sex Robots 'Family Friendly' Deserve to Fail. *Samantha the Sex Robot Has a 'Family Mode'*. London, UK: The New Statesman.

Chan. (2017, October 16). The robotics landscape. *Becoming Human: Artificial Intelligence Magazine*. https://becominghuman.ai/the-robotics-landscape-b371967894c2

Chatzoglou, P. D., Lazaraki, V., Apostolidis, S. D., & Gasteratos, A. C. (2023). Factors affecting acceptance of social robots among prospective users. *International Journal of Social Robotics*. https://doi.org/10.1007/s12369-023-01024-x

Chen, S.-C., Jones, C., & Moyle, W. (2018). Social robots for depression in older adults: A systematic review. *Journal of Nursing Scholarship*, 50, 612622. https://doi.org/10.1111/jnu.12423.

Cheng, L. (2022). Human emotions projected onto androids: A manifestation of internal crisis—human–android interactions in "The Sand-Man" by ETA Hoffmann as example. *Neohelicon*, 49(2), 451–463.

Chuck Wendig: Evolution and ruination. (2015, December). *Locus*, 75, 659.

Cilesiz, S. (2011). A phenomenological approach to experiences with technology: Current state, promise, and future directions for research. *Educational Technology Research and Development*, 59, 487–510.

CLAMP. (2000-2022). Chobits [anime]. *Weekly Young Magazine*. Australia: Madman Entertainment.

Clark, A. (2001). Reasons, robots and the extended mind. *Mind & Language*, 16(2), 121–145.

Clarke, R. (1994, September). Asimov's laws of robotics: Implications for information technology, 2. *IEEE Computer*, 27(1), 57–66. www.rogerclarke.com/SOS/Asimov.html

Coeckelbergh, M. (2011a). Humans, animals, and robots: A phenomenological approach to human-robot relations. *International Journal of Social Robotics*, 3, 197–204.

Coeckelbergh, M. (2011b). You, robot: On the linguistic construction of artificial others. *AI & Society*, 26, 61–69.

Coeckelbergh, M. (2020). *AI Ethics*. Cambridge, MA: MIT Press.

Coles, R. (1991). *The Spiritual Life of Children*. San Francisco, CA: HarperOne.

Columbia 250. *Isaac Asimov*. Retrieved April 30, 2024: https://c250.columbia.edu/

Complete Jewish Bible, The. (Bereshit/Genesis 1:26). *Chabad.org*. Retrieved March 19, 2024: www.chabad.org/library/bible_cdo

Connors, H. L. (2006, February 8). Robot Rosie dispenses meds at Benedictine. *Times-Herald Record*. www.recordonline.com/story/entertainment/local/2006/02/08/robot-rosie-dispenses-meds-at/51118944007/

Context hacking: How to Mess with Art, Media, Law and the Market. (2013, April). Vienna, Austria: mono/monochrom. ISBN-10: 3902796138.

Copley, T. (2007). The power of the storyteller in religious education. *Religious Education*, 102(3), 288–297.

Cox, J. (2010). *An Introduction to the Phenomenology of Religion*. London, UK: A&C Black.

Cox, J. (2023, July 13). AI anxiety: The workers who fear losing their jobs to artificial intelligence. *BBC*. www.bbc.com/worklife/article/20230418-ai-anxiety-artificial-intelligence-replace-jobs

Crain, M. A. (2007). Reconsidering the power of story in religious education. *Religious Education*, 102(3), 241–248.

Crang, M. (1998). *Cultural Geography*. London: Routledge.

Creswell, J. W. (2013). *Qualitative Inquiry and Research Design: Choosing Among Five Approaches* (3rd ed.). Los Angeles: Sage.

Crichton, M. (1973). *Westworld*. London: MGM.

Crist, R. (2017, August 10). Dawn of the sexbots. *CNet*. www.cnet.com/culture/abyss-creations-ai-sex-robots-headed-to-your-bed-and-heart/

Crowe, C. (2005). *Elizabethtown*. Los Angeles, CA: Paramount.

Csikszentmihalyi, M., & Rochberg-Halton, E. (1981). *The Meaning of Things: Domestic Symbols and the Self*. Cambridge: Cambridge University Press.

Cuthbertson, A. (2015, December 9). Virtual reality heaven: How technology is redefining death and the afterlife. *International Business Times*. www.ibtimes.co.uk/virtual-reality-heaven-how-technology-redefining-afterlife-1532429

Da Costa. (2012). *Interview with Johannes Grenzfurthner of monochrom*, Part 3 Archived 17 March 2012 at the Wayback Machine. Furtherfield.

da Mota Pedrosa, A., Näslund, D., & Jasmand, C. (2012). Logistics case study based research: Towards higher quality. *International Journal of Physical Distribution & Logistics Management*, 42(3), 275–295.

Darling, K. (2016, March 8). Extending legal protection to social robots: The effects of anthropomorphism, empathy, and violent behavior towards robotic objects. In *Robot Law* (pp. 213–232). Cheltenham, UK: Edward Elgar Publishing.

Darling, K. (2021). *The New Breed: What Our History with Animals Reveals about Our Future with Robots*. Henry Holt and Company.

Darling, K. (2022, October 26). Robot companions are on their way, but don't worry, they won't replace humans. *BBC ScienceFocus*. www.sciencefocus.com/news/robot-companions-are-on-their-way-but-dont-worry-they-wont-replace-humans

Dautenhahn, K. (2007). Socially intelligent robots: Dimensions of human–robot interaction. *Philosophical Transactions of the Royal Society B: Biological Sciences*, 362(1480), 679–704.

Dautenhahn, K., & Billard, A. (1999, May 1–5). Bringing up robots or—The psychology of socially Intelligent robots: From theory to implementation. In *Proceedings of Third International Conference on Autonomous Agents* (Agents '99). Seattle, WA and Washington, DC: ACM.

Davecat. (2021, August 1820). The life synthetik: My twenty-year journey with artificial companions. In *6th Annual Love and Sex with Robots Conference*.

Davis, E. (2017, November 6). Can robots be Jewish? After Saudi Arabia awards citizenship to a cyborg, rabbis weigh in. *Tablet*. Retrieved May 2, 2024: www.tabletmag.com/sections/news/articles/can-robots-be-jewish

Dawson, T. (2012). Enchantment, possession and the uncanny in ETA Hoffmann's 'The Sandman'. *International Journal of Jungian Studies*, 4(1), 41–54.

De Beauvoir, S. (2023). The second sex. In *Social Theory Re-Wired* (pp. 346–354). London: Routledge.

De Fren, A. (2012). *The Mechanical Bride* [film]. USA: Herwitz.

Dee, S., & Goehrig, G. (1960). *Robot Man*. [Single]. MGM.

Dearnley, C. (2005). A reflection on the use of semi-structured interviews. *Nurse researcher*, 13(1).

DeFalco, A. I. (2016). Beyond prosthetic memory: Posthumanism, embodiment, and caregiving robots. *Age, Culture, Humanities: An Interdisciplinary Journal*, 3.

Dennett, D. C. (1989). *The Intentional Stance*. Cambridge, MA: MIT Press.

Derrida, J. (2002). The animal that therefore I am (more to follow). [Trans. Wills, D.]. *Critical Inquiry*, 28(2), 369–418.

Derry, J. (2013). *Vygotsky: Philosophy and Education*. Hoboken, NJ: John Wiley & Sons.

Devlin, K. (2015). In defence of sex machines: why trying to ban sex robots is wrong. *The Conversation*. Retrieved September 2, 2024: https://theconversation.com/in-defence-of-sex-machines-why-trying-to-ban-sex-robots-is-wrong-47641

Devlin, K. (2018). *Turned on: Science, Sex and Robots*. London, UK: Bloomsbury Publishing.

Dewey, J. (1938). *Experience and Education*. New York: Macmillan.

Dick, P. K. (1968). *Do Androids Dream of Electric Sheep?* New York: Doubleday.

Dick, P. K. (1972). *The Android and the Human*. Vancouver Science Fiction Convention [speech]. Canada: University of British Columbia. https://genius.com/Philip-k-dick-the-android-and-the-human-annotated

Dinkins, S. (2014). *Conversations with Bina48. [Bina48 on racism]*. Retrieved April 15, 2024: www.stephaniedinkins.com/conversations-with-bina48.html

Doll Sweet. (2024). *Sexdoll*. www.dsdoll.us/

Dorman, A. (2023, December 6). This robot comes with snacks: Delivery model is a stress reliever for busy staff, admin says. *McKnight Senior Living*. www.mcknightsseniorliving.com/home/news/tech-daily-news/this-robot-comes-with-snacks-delivery-model-is-a-stress-reliever-for-busy-staff-admin-says/

Drones & Robots. (2024). Cotton Engineering, Texas A&M. https://cottonengineering.tamu.edu/production/drones-robots/

Dudek, S. Y., & Young, J. E. (2022, March 7–10). Fluid sex robots: Looking to the 2LGBTQIA+ community to shape the future of sex robots. In *The Proceedings of HRI 2022*. Sapporo, Hokkaido, Japan.

Duffy, B. R., Rooney, C. F. B., & O'Hare, G. M. (2000). *The Social Robot* [Doctoral dissertation, University College Dublin].

Eaton, M. (2015, February 2). *Evolutionary Humanoid Robotics*. Heidelberg [Germany]: Springer, p. 40. ISBN 9783662445990. OCLC 902724634.

Ebert, R. (2005, July). *Roger Tribute* [Video & transcript; 4 April, 2014]. Chicago, IL. www.rogerebert.com/empathy/video-roger-ebert-on-empathy

Edwards, J. (2017, November 13). An interview with the artificially intelligent robot Sophia. *Business Insider*. Retrieved August 24, 2024: https://www.weforum.org/agenda/2017/11/an-interview-with-the-artificially-intelligent-robot-sophia/

Ehrsson, H. H. (2007). The experimental induction of out-of-body experiences. *Science*, 317(5841), 1048. https://doi.org/10.1126/science.1142175

Ehrsson, H. H., Holmes, N. P., & Passingham, R. E. (2007). Touching a rubber hand: Feeling of body ownership is associated with activity in multisensory brain areas. *The Journal of Neuroscience*, 27(34), 10564–10573. https://doi.org/10.1523/jneurosci.0800-07.2007

Eichrodt, W. (1967). The "soul" (nepeš). In *Eichrodt, Theology of the Old Testament* (2 vols., pp. 2:131–45, at 136). London: SCM Press.

Embree, L., & Nenon, T. (Eds.). (2012). *Husserl's Ideen* (Vol. 66). Philadelphia, PA: Springer Science & Business Media.

England, R. (2020, June 26). AI robot 'Erica' will star in $70 million sci-fi movie 'b'. *Endgadget*. www.engadget.com/ai-robot-erica-will-star-in-70-million-scifi-movie-b-130539023.html

Erica: Man Made. (2017). Caligareanu, I [documentary]. London: Guardian Documentaries. www.facebook.com/theguardian/videos/erica-man-made/10156243103311323/Estrada, D. (2018, June 17). Sophia and her critics: The ethics of human likeness—Part 1. *Medium*. https://medium.com/@eripsa/sophia-and-her-critics-5bd22d859b9c

Estrada, D. (2020, May). Human supremacy as posthuman risk. *The Journal of Sociotechnical Critique*, 1(1), 1–40. https://doi.org/10.25779/j5ps-dy87

Evans, S. (2016, December 23). This woman Is determined to marry this robot. *Men's Health*. Retrieved May 19, 2024: www.menshealth.com/trending-news/a19535003/woman-determined-to-marry-robot/

Evjemo, L. D., Gjerstad, T., Grøtli, E. I., & Sziebig, G. (2020). Trends in smart manufacturing: Role of humans and industrial robots in smart factories. *Current Robotics Reports*, 1, 35–41.

Eyssel, F., Kuchenbrandt, D., Bobinger, S., De Ruiter, L., & Hegel, F. (2012, March). If you sound like me, you must be more human: On the interplay of robot and user features on human-robot acceptance and anthropomorphism. *In Proceedings of the Seventh Annual ACM/IEEE International Conference on Human-Robot Interaction* (pp. 125–126).

Fernandes, A. T. (2009). *Hiding Hiroshima: Blowing Away Atomic Responsibility* [Master's thesis, USA: Rhode Island College], 28. https://digitalcommons.ric.edu/etd/28

Fink, J., Mubin, O., Kaplan, F., & Dillenbourg, P. (2012, May). Anthropomorphic language in online forums about Roomba, AIBO and the iPad. In *2012 IEEE Workshop on Advanced Robotics and Its Social Impacts (ARSO)* (pp. 54–59). Munich, Germany, Piscataway, NJ: IEEE.

Fiorini, L., De Mul, M., Fabbricotti, I., Limosani, R., Vitanza, A., D'Onofrio, G., Tsui, M., Sancarlo, D., Giuliani, F., Greco, A., Guiot, D., Senges, E., & Cavallo, F. (2021). Assistive robots to improve the independent living of older persons: Results from a needs study. *Disability and Rehabilitation: Assistive Technology*, 16(1), 92–102. https://doi.org/10.1080/17483107.2019.1642392

Fischer, L. (2003). *Designing Women: Cinema, Art Deco, and the Female Form*. New York, NY: Columbia University Press.

Flores, I. (2024, March 24). Saudi Arabia's first humanoid robot 'Sara' embodies nation's values in design. *TechTimes*. www.techtimes.com/articles/302898/20240324/saudi-arabias-first-humanoid-robot-sara-embodies-nations-values-design.htm

Fong, T., Nourbakhsh, I., & Dautenhahn, K. (2003). A survey of socially interactive robots. *Robotics and Autonomous Systems*, 42(3–4), 143–166.

Forbes, B. (1975). *The Stepford Wives* [film]. Los Angeles, CA: Palomar Pictures, Intl.

Forden, J., Radpour, S., Conway, E., & Ghilarducci, T. (2023, April). *Reducing the Unequal Burden of Eldercare Work*. New York, NY: The New School, Schwartz Center for Economic Policy Analysis.

Förster, Y. (2023). Technology and the spiritual: From prayer bots to the Singularity. In N. R. B. Loewen & A. Rostalska (Eds.), *Diversifying Philosophy of Religion: Critiques, Methods and Case Studies* (pp. 263–278). London: Bloomsbury.

Foucault, M. (1978). *The History of Sexuality, Volume 1: An Introduction*. New York: Pantheon Books.

Frack, E., & Chapman, J. (2021). The unique experience s of the sandwich generation. *Sociological Viewpoints*, 35(1).

Frank, A. W. (2004). Narratives of Spirituality and Religion in End-of-Life Care. In *Narrative Research in Health and Illness* (pp. 132–145). Hoboken, NJ: Wiley. https://doi.org/10.1002/9780470755167

Fraser, D. (2008). Understanding animal welfare. *Acta Veterinaria Scandinavica*, 50(1), 1–7.

Friesinger, G., Grenzfurthner, J., Schneider, F. A., Ballhausen, T., & Fink, R. (2013). *Context Hacking: How to Mess with Art, Media, Law and the Market.* Vienna, Austria: PLUS Research.

Freud, S. (1923). The Ego and the Id. In J. Strachey et al. (Trans.), *The Standard Edition of the Complete Psychological Works of Sigmund Freud* (p. XIX). London: Hogarth Press.

Fullerton, H. (2019, July 25). Rutger Hauer dissects his iconic 'tears in rain' Blade Runner monologue. *RadioTimes.* www.radiotimes.com/tv/sci-fi/blade-runner-tears-in-rain-speech/

Gallagher, S. (2006). *How the Body Shapes the Mind.* Oxford, UK: Clarendon Press.

Gallagher, S., & Zahavi, D. (2023). Phenomenological approaches to self-consciousness. In Edward N. Zalta & Uri Nodelman (Eds.), *Stanford Encyclopedia of Philosophy* (Winter 2023 Edition). Archived from the original on 11 July 2023. Stanford, CA: Stanford Encyclopedia of Philosophy. Retrieved July 31, 2023: https://plato.stanford.edu/archives/win2023/entries/self-consciousness-phenomenological/

Galván, F. (1999). On belonging and not belonging: A conversation with Amit Chaudhuri. *Wasafiri*, 15, 42–50.

Gammelgaard, B. (2017). The qualitative case study. *The International Journal of Logistics Management*, 28(4), 910–913.

Ganpati Puja with the Scientific Involvement of Robotic Arm at Monarch Innovation. (2021, September 17). *Monarch Innovations.* www.monarch-innovation.com/ganesh-aarti-with-robotic-arm-technology

Garland, A. (2014). *Ex Machina* [Film]. A24 Films.

Gee, J. P. (1999). Critical issues: Reading and the new literacy studies: Reframing the national academy of sciences report on reading. *Journal of Literacy Research*, 31(3), 355–374.

Geertz, C. (1973). *The Interpretation of Cultures: Selected Essays.* New York: Basic Books.

Gemmel, A. (2018, June 24). One of the world's most famous sex robots can now revoke her consent. *Dazed.* Retrieved August 20, 2024: https://www.dazeddigital.com/science-tech/article/40478/1/samantha-the-sex-robot-can-refuse-sex-if-not-in-the-mood

Gencarelli, M. (2013, January 19). Rachael Ma talks about her role in 'Robot & Frank'. *MediaMikes.* https://mediamikes.com/2013/01/rachael-ma-talks-about-her-role-in-robot-frank/

Gergen, K. J. (1995, April). Social construction and the transformation of identity politics. In *New School for Social Research Symposium* (Vol. 7). New York, NY: Routledge.

Gergen, K. J., & Gergen, M. (2004). *Social Construction: Entering the Dialogue.* Chagrin Falls, OH: Taos Institute.

Giarratano, J. C., & Riley, G. (2005). *Expert Systems: Principles and Programming* (4th ed.). Clifton Park, NY: Course Technology.

Giddens, A. (1984). *The Constitution of Society: Outline of the Theory of Structuration.* Oakland, CA: University of California Press.

Giger, J. C., Moura, D., Almeida, N., & Piçarra, N. (2017, May). Attitudes towards social robots: The role of gender, belief in human nature uniqueness, religiousness and interest in science fiction. In *Proceedings of II International Congress on Interdisciplinarity in Social and Human Sciences* (Vol. 11, p. 509). Faro, PT: CIEO – Research Centre for Spatial and Organizational Dynamics.

Gilbert, J. (2020, October 26). Rosie the Root Robot. *MrsGilbertRocks.com*. www.msgilbertrocks.com/blog/2020/8/11/rosie-the-root-robot

Gill, P., Stewart, K., Treasure, E., & Chadwick, B. (2008). Methods of data collection. In *Qualitative Research: Interviews and Focus Groups* (pp. 291–295). London, UK: British Dental Journal. https://doi.org/10.1038/bdj.2008.192

Gillespie, C. (2007). *Lars and the Real Girl*. Metro-Goldwyn-Mayer (MGM).

Glaser, A. (2016, April 4). The Scarlett Johansson bot Is the robotic future of objectifying women. *WIRED*. www.wired.com/2016/04/the-scarlett-johansson-bot-signals-some-icky-things-about-our-future/

Gleisner, J. (2017, November). Robots, race, and algorithms: Stephanie Dinkins at Recess Assembly. *Art21 Magazine*. http://magazine.art21.org/2017/11/07/robotsrace-and-algorithms-stephanie-dinkins-atrecess-assembly/#.YTCt99MzZN0

Glinert, L. (2001, January). Golem! The making of a modern myth. In *Symposium: A Quarterly Journal in Modern Literatures* (Vol. 55, No. 2, pp. 78–94). Taylor & Francis Group.

Goertzel, B. (2010). Toward a formal definition of real-world general intelligence. In *Proceedings of AGI* (p. 10). Dordrecht, NL: Atlantis Press. https://doi.org/10.2991/agi.2010.17

Goetz, J., Kiesler, S., & Powers, A. (2003, November). Matching robot appearance and behavior to tasks to improve human-robot cooperation. In *The 12th IEEE International Workshop on Robot and Human Interactive Communication, 2003. Proceedings. ROMAN 2003* (pp. 55–60), Millbrae, CA, USA, New York City, NY: IEEE.

Goffman, E. (2002). *The Presentation of Self in Everyday Life, 1959* (p. 259). New York, NY: The Overlook Press.

Gould, H., Arnold, M., Kohn, T., Nansen, B., & Gibbs, M. (2021). Robot death care: A study of funerary practice. *International Journal of Cultural Studies*, 24(4), 603–621.

Goyal, N. (2017, February 6). This Woman is in Love with Her 3D Printed Robot and Now She Wants to Marry It. *Industry Tap into News*. Retrieved August 24, 2024: https://www.industrytap.com/woman-love-3d-printed-robot-now-wants-marry/40688

Griffith, E. (2018, May 15). Henry the sexbot wants to know all your hopes and dreams. *WIRED*. www.wired.com/story/henry-the-sexbot-wants-to-know-all-your-hopes-and-dreams/

Groopman, J. (2009, November 2). Robots that care: Advances in technological therapy. *The New Yorker*. https://archives.newyorker.com/newyorker/2009-11-02/flipbook/066/

Guizzo, E. (2010, April 22). A Japanese roboticist is building androids to understand humans—starting with himself. *IEEE Spectrum*. New York, NY.

Guizzo, E. (2014, July 16). Cynthia Breazeal unveils Jibo, a social robot for the home. *IEEE Spectrum*. https://spectrum.ieee.org/cynthia-breazeal-unveils-jibo-a-social-robot-for-the-home

Gunkel, D. J. (2018). *Robot Rights*. Cambridge, MA: MIT Press.

Guo, W. (2021). Positronic brain vs human brain: Robots' brain as the representation of ethical guiding principles in Isaac Asimov's the Complete Robot. *Kritika Kultura*, 37.

Gusfield, J. R., & Michalowicz, J. (1984). Secular symbolism: Studies of ritual, ceremony, and the symbolic order in modern life. *Annual Review of Sociology*, 10, 417–435.

Hajime Sorayama's Science Frisson: Sci-fi's preeminent romantic releases his first compendium. (2010, May 21). *Nowness*. London, UK. Retrieved September 3, 2024: www.nowness.com/story/hajime-sorayamas-science-frisson

Hall, S. (1997). *Representation: Cultural Representations and Signifying Practices*. Washington, DC: Sage.

Hamill, J. (2017, November 6). Samantha the sex robot could be 'mass produced' in Wales, creators claim. *The Sun*. www.thesun.co.uk/tech/4851058/samantha-the-sex-robot-could-be-mass-produced-in-wales-creators-claim/

Hanna-Barbera Productions. (1962 September–1963 March). *The Jetsons* [Television show]. ABC.

Hanson Robotics. (2024a). *Bina: Custom Character Robot*. www.hansonrobotics.com/bina48-9/

Hanson Robotics. (2024b). *Philip K. Dick Research Robot*. https://www.hansonrobotics.com/philip-k-dick/

Haraway, D. (2013a). A cyborg manifesto: Science, technology, and socialist-feminism in the late twentieth century. In *The Transgender Studies Reader* (pp. 103–118). UK: Routledge.

Haraway, D. (2013b). *Simians, Cyborgs, and Women: The Reinvention of Nature*. New York: Routledge.

Harris, M. (2011). *Cows, Pigs, Wars, and Witches: The Riddles of Culture*. New York: Vintage Books.

Harvey, M. E. (2022). *Let's Talk: A Rabbi Speaks to Christians*. New York: Teach Me Judaism Press.

Harvey, P. (2012). *An Introduction to Buddhism: Teachings, History and Practices*. Cambridge: Cambridge University Press.

Heerink, M., Kröse, B., Evers, V., & Wielinga, B. (2010). Assessing acceptance of assistive social agent technology by older adults: the Almere Model. *International Journal of Social Robotics*, 2, 361–375. https://doi.org/10.1007/s12369-010-0068-5

Heidegger, M. (1962). *Being and Time*. Oxford: Basil Blackwell.

Henderson, M. (2007, April 24). Human rights for robots? We're getting carried away. *The Times*. www.thetimes.co.uk/article/human-rights-for-robots-were-getting-carried-away-xfbdkpgwn0v

Henry, B. (2019). From Golem to Cyborg. Symbolic reconfigurations of an ancient Monstrum. In *Monsters, Monstrosities and the Monstrous in Culture and Society* (Vol. 1, pp. 233–256). Vernon Press.

Hernandez, V. (2017, May 13). Sex doll brothel in Barcelona moves to UK after hookers' protest. *BlastingNews*. Retrieved June 5, 2024: https://us.blastingnews.com/lifestyle/2017/05/sex-doll-brothel-in-barcelona-moves-to-uk-after-hookers-protest-001696603.html

Herzog, H. (2011). *Some We Love, Some We Hate, Some We Eat*. New York: Harper.

"Hey, Clockmaker". (2015). *Mishkan Hanefesh: Machzor for the Days of Awe (Vol. Yom Kippur)* (p. 177). New York: CCAR Press.

Hirai, K., Hirose, M., Haikawa, Y., & Takenaka, T. (1998, May). The development of Honda humanoid robot. In *Proceedings. 1998 IEEE International Conference on Robotics and Automation (Cat. No. 98CH36146)* (Vol. 2, pp. 1321–1326). IEEE.

Hoffmann, E. T. A. (1816). *The Sandman*. Transl. John Oxenford. chrome-extension://efaidnbmnnnibpcajpcglclefindmkaj/www.ux1.eiu.edu/~rlbeebe/sandman.pdf

Hoffman, G. (2019, May 1). Anki, Jibo, and Kuri: What we can learn from social robots that didn't make It It's been a tough few years for social home robots: Where do we go from here? *IEEE Spectrum*. https://spectrum.ieee.org/anki-jibo-and-kuri-what-we-can-learn-from-social-robotics-failures

Hoffman, G., & Ju, W. (2014). Designing robots with movement in mind. *Journal of Human-Robot Interaction*, 3(1), 91–122.

Hogan, J. (2009). The comic book as symbolic environment: The case of Iron Man. *ETC: A Review of General Semantics*, 66(2), 199–214.

Honda. (2002). "The power of dreams." [print advertisement]. Private collection, Julie Carpenter. San Francisco, CA, USA.

Honda in America. (2019, January 10). *Honda Walking Assist Device Receives Clearance from U.S. Food and Drug Administration*. Rayomnd, OH, USA. https://hondainamerica.com/news/honda-walking-assist-device-receives-clearance-from-u-s-food-and-drug-administration

Honda News. (2005, June 1). Mickey Welcomes ASIMO To Disneyland's 50th Anniversary. https://global.honda/en/newsroom/worldnews/2005/c050601b.html

Honda News. (2010, January 22). *Honda Premieres 'Living with Robots' Short Film Documentary at Sundance Film Festival*. https://hondanews.com/en-US/honda-corporate/releases/release-e2d03d45d6c0a32ee025d9004c34ba4b-honda-premieres-living-with-robots-short-film-documentary-at-sundance-film-festival

Honda Robotics. (2024). *What We Learned from ASIMO*. https://global.honda/en/robotics/asimo/

Hornyak, T. (2014, June 5). Meet Pepper, the 'love-powered' humanoid robot that knows how you're feeling. *PCWorld*. www.pcworld.com/article/439658/softbanks-humanoid-robot-pepper-knows-how-youre-feeling.html

Horstmann, A. C., & Krämer, N. C. (2020). Expectations vs. actual behavior of a social robot: An experimental investigation of the effects of a social robot's interaction skill level and its expected future role on people's evaluations. *PLoS ONE*, 15(8), e0238133.

Husserl, E. (1962). *Ideas: General Introduction to Pure Phenomenology.* Trans. by W. F. Boyce Gibson. New York: Collier Books.

Husserl, E., & Moran, D. (2012). *Ideas: General Introduction to Pure Phenomenology.* London: Routledge. https://doi.org/10.4324/9780203120330.

iRobot Blog. (2022, June 1). *The Top 50 Most Popular Roomba Names.* Retrieved February 8, 2024: https://blog.irobot.com/roomba-names/

Istvan, Z. (2016, January 13). I visited a community where people upload their personalities to 'mindfiles' so they can live on after death. *Business Insider.* Retrieved April 26, 2024: www.businessinsider.com/a-visit-to-martine-rothblatts-terasem-community-2016-1

Jackson, J. C., & Yam, K. C. (2023, July 25). The in-credible robot priest and the limits of robot workers. *Scientific American.* Retrieved March 19, 2024: www.scientificamerican.com/article/the-in-credible-robot-priest-and-the-limits-of-robot-workers/

Jackson, J. C., Yam, K. C., Tang, P. M., Liu, T., & Shariff, A. (2023). Exposure to robot preachers undermines religious commitment. *Journal of Experimental Psychology: General*, 152(12), 3344–3358. https://doi.org/10.1037/xge0001443

Jecker, N. S. (2020, November 16). Nothing to be ashamed of: Sex robots for older adults with disabilities. *Journal of Medical Ethics*, 47, 26–32. https://doi.org/10.1136/medethics-2020-106645

Jecker, N. S. (2021, January 17). Sex robots for older adults with disabilities: Reply to critics. *Journal of Medical Ethics*, 113. https://doi.org/10.1136/medethics-2020-107148

Jeffs, R., & Blackwood, G. (2016). Whose real? Encountering new frontiers in Westworld. *Media Peripheries-Situated in Aotearoa, Regional in Focus, Global in Scope*, 16(2).

Jensen, C. B., & Blok, A. (2013). Techno-animism in Japan: Shinto cosmograms, actor-network theory, and the enabling powers of non-human agencies. *Theory, Culture & Society*, 30(2), 84–115.

Jha, A. (2004, February 17). Meet the home help of the future. *The Guardian.* Retrieved May 9, 2024: www.theguardian.com/science/2004/feb/17/sciencenews.uk

Jibo, the First Social Robot for the Home, Named One of TIME's Best Inventions of 2017. (2017, November 16). PRNewswire.com. Retrieved February 1, 2024: www.prnewswire.com/news-releases/jibo-the-first-social-robot-for-the-home-named-one-of-times-best-inventions-of-2017-300557729.html

Jones, C. Fears of employee displacement as Amazon brings robots into warehouses. *The Guardian.* Retrieved April 12, 2024: www.theguardian.com/technology/2023/oct/18/amazon-robot-warehouses-digit-workers

Jonze, S. (2013). *Her.* Burbank, CA: Warner Bros Pictures.

Jøranson, N., Pedersen, I., Rokstad, A. M., Aamodt, G., Olsen, C., & Ihlebæk, C. (2016, August 28). Group activity with Paro in nursing homes: systematic investigation of behaviors in participants. *International Psychogeriatrics*, 8, 1345–1354. https://doi.org/10.1017/S1041610216000120. PMID: 27019225.

Jörgensen, A. (2016). *Speaking of Gender: A Sociolinguistic Exploration of Voice and Transgender Identity* [Master's thesis, University of Copenhagen].

Kageki, N. (2012, June 12). An uncanny mind: Masahiro Mori on the Uncanny Valley and beyond. *IEEE Spectrum.* Retrieved March 14, 2024: https://spectrum.ieee.org/an-uncanny-mind-masahiro-mori-on-the-uncanny-valley

Kagitcibasi, C. (2017). Doing psychology with a cultural lens: A half-century journey. *Perspectives on Psychological Science*, 12(5), 824–832.

Kahn, P. H., Freier, N. G., Friedman, B., Severson, R. L., & Feldman, E. N. (2004, September). Social and moral relationships with robotic others? In *RO-MAN 2004. 13th IEEE International Workshop on Robot and Human Interactive Communication* (IEEE Catalog No. 04TH8759) (pp. 545–550), Kurashiki, Japan and New York: IEEE.

Kakubayashi, M. (2021, March). 'Reality is catching up to science fiction': An interview with Hiroshi Ishiguro. *KULTURTECHNIKEN 4.0 Goethe-Institut*. Retrieved February 17, 2024: www.goethe.de/prj/k40/en/fil/ish.html

Kastner, J. (2024, January 4). Will robots replace garment workers in future? *JustStyle*. Retrieved April 12, 2024: www.just-style.com/features/will-robots-replace-garment-workers-in-future/

Katsikopoulou, M. (2001, December 9). Hajime Sorayama reunites with Heavy Metal magazine for Sexy Robot cover series. *Designboom*. Retrieved May 6, 2024: www.designboom.com/art/heavy-metal-magazine-hajime-sorayama-sexy-robot-cover-series-12-08-2021/

Kavner, L. (2012, July 9). You, Robot. *HuffPost: Tech*. Retrieved April 13, 2013: www.huffingtonpost.com/2012/07/17/you-robot-personal-robots_n_1660362.html

Kedar, H. E. (2018). *The Thomann Intervention* [documentary]. Retrieved August 24, 2024: https://www.youtube.com/watch?v=KKh1RojOyQI

Kertész, C., & Turunen, M. (2017, November 22–24). What can we learn from the long-term users of a social robot? *Social Robotics: 9th International Conference, ICSR 2017, Proceedings 9* (pp. 657–665). Tsukuba, Japan: Springer.

Khaleghipour, M., & Shahghasemi, E. (2020, September 4–6). Fantasy animations and children's imagination: A qualitative study. *2nd International Conference on Future of Social Sciences and Humanities*. Prague, Czech. Retrieved January 2024: www.dpublication.com/wp-content/uploads/2019/09/451-51.pdf

Kibo Robot Project. (2015). Story and Report (Updated on November 17, 2015). Retrieved September 5, 2024: https://kibo-robo.jp/

Kidd, C. D., & Breazeal, C. (2008, September). Robots at home: Understanding long-term human-robot interaction. In *2008 IEEE/RSJ International Conference on Intelligent Robots and Systems* (pp. 3230–3235). IEEE.

Kidd, C. D., Taggart, W., & Turkle, S. (2006, May). A sociable robot to encourage social interaction among the elderly. In *Proceedings 2006 IEEE International Conference on Robotics and Automation*. ICRA 2006 (pp. 3972–3976), Orlando, FL and New York: IEEE.

Kiesler, S., Powers, A., Fussell, S. R., & Torrey, C. (2008). Anthropomorphic interactions with a robot and robot–like agent. *Social Cognition*, 26(2), 169–181.

Kim, M. S., & Kim, E. J. (2013). Humanoid robots as the cultural other: Are we able to love our creations? *AI & Society*, 28, 309–318.

King James Bible. (2024). Retrieved online February 10, 2024: www.kingjames-bibleonline.org/

Kinyon, K. (1999). The phenomenology of robots: Confrontations with death in Karel Čapek's "RUR". *Science Fiction Studies*, 26, 379–400.

Kishiro, Y. (1990–1995). Battle: Angel Alita (Gunnm). *Business Jump Magazine*. Tokyo, Japan: Shueisha, Inc.

Kit, B. (2022, May 26). Jude law to star in Jon Watts' 'Star Wars' series. *The Hollywood Reporter*. hollywoodreporter.com

Kivel, P. (2009). About Christian hegemony. *Challenging Christian Hegemony*. http://christianhegemony.org/

Kobayashi, E. (2007, February 6). How to be buried alive. *Toronto Star*.

Koistinen, A. K. (2016). The (care) robot in science fiction: A monster or a tool for the future? *Confero: Essays on Education, Philosophy and Politics*, 4(2), 97–109.

Korstanje, M. E. (2024). AI and robots in science fiction movies: Why should we trust in AI? In *AI and Emotions in Digital Society* (pp. 141–153). IGI Global.

Kroehling, R. (2009). *2B: The Era Flesh Is Over*. Bristol, VT: Terasem Media and Films.

Kurzweil, R. (2005). *The Singularity Is Near*. New York: Viking Press.

Kwon, M., Jung, M. F., & Knepper, R. A. (2016, March). Human expectations of social robots. In *2016 11th ACM/IEEE International Conference on Human-Robot Interaction* (pp. 463–464), Christchurch, New Zealand and New York: IEEE.

Lacan, J. (1953). Some reflections on the ego. *The International Journal of Psycho-Analysis*, 34, 11.

Lacan, J. (2004). The mirror stage as formative of the function of the I. In reading French psychoanalysis. In J. Rivkin & M. Ryan (Eds.), *Literary Theory: An Anthology* (2nd ed., pp. 97–104). Malden, MA: Blackwell Publishing.

Lang, F. (1927). *Metropolis* [film]. Los Angeles, CA: Paramount Pictures.

Lapin, A. (2022, May 14). A rabbi who 'speaks to Christians' condemned them on Twitter. It cost him his job. *The Times of Israel*. Retrieved September 4, 2024: https://www.timesofisrael.com/a-rabbi-who-speaks-to-christians-condemned-them-on-twitter-it-cost-him-his-job/

Lazar, A., Edasis, C., & Piper, A. M. (2017, May). Supporting people with dementia in digital social sharing. In *Proceedings of the 2017 CHI Conference on Human Factors in Computing Systems, Denver, CO* (pp. 2149–2162). New York, NY: ACM. https://doi.org/10.1145/3025453.3025586

Leane, R. (2017, July 4). Spider-Man: Homecoming—director Jon Watts interview. *Den of Geek*. denofgeek.com

Lechner, M. (2008, March 4). Pop Guérilla. *Libération*. www.ecrans.fr

Lee, S., Lieber, L., Heck, D., & Kirby, J. (1963). *Tales of Suspense* no. 39. US: Marvel Comics.

Lelièvre, M., Zebrowski, R., & Gressier Soudan, E. (2022, May). Robots and choreography: A contribution to artificial sentience characterization. In *Advanced Information Systems Engineering Workshops: CAiSE 2022 International Workshops, Leuven, Belgium, June 6–10, 2022, Proceedings* (pp. 93–102). Cham: Springer International Publishing.

Levin, I. (1972). *The Stepford Wives*. New York: Random House.

Levinson, K. (2024, March 5). States get some help to recruit caregivers for an aging population. *Route50*. www.route-fifty.com/workforce/2024/03/states-get-some-help-recruit-caregivers-aging-population/394707

Levy, D. (2008). *Love and Sex with Robots*. New York: Harper Perennial.

Liberman-Pincu, E., Parmet, Y., & Oron-Gilad, T. (2023). Judging a socially assistive robot by its cover: The effect of body structure, outline, and color on users' perception. *ACM Transactions on Human-Robot Interaction*, 12(2), 1–26.

Linder, C. (2020, June 25). This AI robot just nabbed the lead role in a sci-fi movie. *Popular Mechanics*. www.popularmechanics.com/technology/robots/a32968811/artificial-intelligence-robot-movie-star-erica/

Löffler, D., Luthe, S., Hurtienne, J., & Nord, I. (2020). From experiential to existential questions: An interdisciplinary view on social robots in religious settings. In *Artificial Intelligence* (pp. 293–305). Leiden, The Netherlands: Brill. https:///doi.org/10.30965/9783957437488_019

Lopez, I. F. H. (1991). The social construction of race. In Julie Rivkin & Michael Ryan (Eds.), *Literary Theory: An Anthology* (2nd ed., pp. 964–974). Maiden, MA: Blackwell Publishing.

Lorber, J. (2021). *The New Gender Paradox: Fragmentation and Persistence of the Binary*. San Francisco, CA: John Wiley & Sons.

Lorber, J., & Farrell, S. A. (Eds.). (1991). *The Social Construction of Gender* (pp. 309–321). Newbury Park, CA: Sage.

Lucas, G. (1977). *Star Wars: IV: A New Hope*. Los Angeles, CA: 20th Century Fox.

Luck, L., Jackson, D., & Usher, K. (2006). Case study: A bridge across the paradigms. *Nursing Inquiry*, 13(2), 103–109.

MacDorman, K. F., & Ishiguro, H. (2006). The uncanny advantage of using androids in cognitive and social science research. *Interaction Studies*, 7(3), 297–337.

MacWilliam, S. (2023). Playthings and corpses—turning women into dead body objects: Sexual objectification, victimisation, representation and consent in art and sex dolls/robots. In K. Richardson & C. Odlind (Eds.), *Man-Made Women. Social and Cultural Studies of Robots and AI*. Cham: Palgrave Macmillan. https://doi.org/10.1007/978-3-031-19381-1_5

Magaldi, D., & Berler, M. (2020). Semi-structured interviews. In *Encyclopedia of Personality and Individual Differences* (pp. 4825–4830).

Making Mr. Right. (1987). Seidelman, S. [Film]. Orion Pictures.

Maller, A. S. (2024, April 2). Robots are performing Hindu rituals and may replace clergy. *MuslimMirror*. https://muslimmirror.com/eng/robots-are-performing-hindu-rituals-and-may-replace-clergy/

Mantio, B. (1984, May 8). *The Transformers (01)*. New York, NY: Marvel Comics.

Mar, A. (17 October, 2017). *Are we ready for intimacy with androids? Love in the time of robots*. San Francisco, CA: WIRED.

Matt McMullen interview for 'The sex robots are coming'. (2017, November 14). [transcript]. Channel4. www.channel4.com/press/news/matt-mcmullen-interview-sex-robots-are-coming

Matyszcyk, C. (2015, September 22). Buyers of emotional robot must agree not to have sex with it, report says. *CNet*. www.cnet.com/culture/buyers-of-emotional-robot-must-agree-not-to-have-sex-with-it-report-says/

Mayor, A. (2018). *Gods and robots: Myths, machines, and ancient dreams of technology.* Princeton, NJ: Princeton University Press.

McCurry, J. (2015, December 31). Erica, the 'most beautiful and intelligent' android, leads Japan's robot revolution. *The Guardian.* www.theguardian.com/technology/2015/dec/31/erica-the-most-beautiful-and-intelligent-android-ever-leads-japans-robot-revolution

McFague, S. (1982). *Metaphorical Theology: Models of God in Religious Language.* Minneapolis, MN: Fortress Press.

McGowan, K. (2024, February). *Love, Dementia, and Robots* (pp. 62–73). San Francisco, CA: Condé Nast. WIRED.

McLuhan, M. (1959). Myth and mass media. *Daedalus,* 88(2), 339–348.

McLuhan, M. (1964). *Understanding Media: The Extensions of Man.* London: Routledge.

McLuhan, M., & Fiore, Q. (1967). *The Medium Is the Massage: An Inventory of Effects.* New York: Random House.

Medeiros, M. (2018). Where do our sex dolls go after we die? *Medium.* https://onezero.medium.com/where-do-our-sex-bots-go-after-we-die-ddf336e458a8

Merleau-Ponty, M. (1962). *Phenomenology of Perception.* London: Routledge & Kegan Paul.

Mickey welcomes ASIMO to Disneyland's 50th anniversary. (2005, June 2). *Physorg.* https://phys.org/news/2005-06-mickey-asimo-disneyland-50th-anniversary.html

Mičulková, D. (2016). *A Comparison of American Comics and Japanese Manga: Superman vs. Astro Boy* [Thesis, Tomas Bara University in Zlín].

Miller, D. A. (1981). The 'sandwich' generation: Adult children of the aging. *Social Work,* 26(5), 419–423.

Miraikan Museum. (2022, March 18–31). *Thank You ASIMO!—Congratulations on Graduating from Miraikan!* Tokyo, Japan. www.miraikan.jst.go.jp/en/events/202203182359.html

Mishkan haNefesh: Machzor for the Days of Awe. (2015). Eds. E. Goldberg, J. Marder, S. Marder & L. Morris (vol. Yom Kippur, p. 177). New York: CCAR Press.

Mogg, T. (2015, September 8). Man arrested for assaulting Pepper, the robot that can read your emotions. *Digital Trends.* www.digitaltrends.com/cool-tech/man-arrested-for-assaulting-pepper-the-robot-that-can-read-your-emotions/

Mohamed, P. (2022, July 30). All that burns unseen. *SLATE.* https://slate.com/technology/2022/07/all-that-burns-unseen-premee-mohamed.html

More than 1 in 10 people in Japan are aged 80 or over. Here's how its ageing population is reshaping the country. (2023, September 28). [report]. *Web Economic Forum.* www.weforum.org/agenda/2023/09/elderly-oldest-population-world-japan/

Morin, R. (2014, February 3). Silicone love: Davecat's Life with a synthetic wife and mistress. *VICE.* www.vice.com/en/article/znwnpw/silicone-love-davecats-life-with-his-synthetic-wife-and-mistress

Moye, D. (2017, September 29). Sex robot molested at electronics festival, creators say. *HuffingtonPost.* www.huffpost.com/entry/samantha-sex-robot-molested_n_59cec9f9e4b06791bb10a268

Mulvey, L. (2013). Visual pleasure and narrative cinema. In *Feminism and Film Theory* (pp. 57–68). UK: Routledge.

Nadine Social Robot. *MiraLab*. Geneva, SU: University of Geneva. Retrieved August 25, 2024: https://www.miralab.ch/index.php/nadine-social-robot/

Nakatani, H. (2019, October 31). Population aging in Japan: Policy transformation, sustainable development goals, universal health coverage, and social determinates of health. *Global Health & Medicine*, 1(1), 3–10. https://doi.org/10.35772/ghm.2019.01011. PMID: 33330747

Narvaez, D. (2016). *Embodied Morality: Protectionism, Engagement and Imagination.* Berlin, Germany: Springer.

Nasr, S. H. (2006). *Islamic Philosophy from Its Origin to the Present: Philosophy in the Land of Prophecy.* Albany, NY: State University of New York Press.

Nass, C. I., & Brave, S. (2005). *Wired for Speech: How Voice Activates and Advances the Human-Computer Relationship.* Cambridge, MA: MIT Press.

Nass, C., & Moon, Y. (2000). Machines and mindlessness: Social responses to computers. *Journal of Social Issues*, 56(1), 81–103.

National Institute of Advanced Industrial Science and Technology (AIST). (2004, December 8). *Seal-Type Robot "Paro" to Be Marketed with Best Healing Effect in the World* (Trans. 2004, Sept 17). www.aist.go.jp/aist_e/list/latest_research/2004/20041208_2/20041208_2.html

National Institute of Advanced Industrial Science and Technology (AIST). AIST Stories. (2014, No. 3). *The World's Only Therapeutic Robot: Paro the Robot Seal Is Active in Medical Treatment and Welfare.* www.aist.go.jp/index_en.html

Navon, M. (2023). Let us make man in our image-a Jewish ethical perspective on creating conscious robots. *AI and Ethics*, 1–16

Neon Genesis Evangelion. (1995, October). Eds. H. Anno, Mazayuki & K. Tsurumaki. Tokyo: Tatsunoko Prod.

Neuhuber, J. (2024). *Je suis auto.* Austria: Monochrom.

Newitz, A. (2017). *Autonomous,* New York, NY: Tor Books.

Nichols, G. (2018, January 22). Robot fired from grocery store for utter incompetence. *ZDNet.* www.zdnet.com/article/robot-fired-from-grocery-store-for-utter-incompetence/

Nishio, S., Ishiguro, H., & Hagita, N. (2007, June). Geminoid: Teleoperated android of an existing person. In A. De Pino Filho (Ed.), *Humanoid Robots: New Developments.* London, UK: InTech Open. https://doi.org/10.5772/4876. Retrieved September 2, 2024: https://www.intechopen.com/chapters/240

Nocks, L. (1998). The Golem: Between the technological and the divine. *Journal of Social and Evolutionary Systems*, 21(3), 281–303.

Nomura, T. (2017). Robots and gender. *Gender and the Genome*, 1(1), 18–25.

Nowness. (2010, May 21). *Hajime Sorayama's Science Frisson.* www.nowness.com/story/hajime-sorayamas-science-frisson

Nungesser, S. (2021, April 8). Sophia the robot becomes first robot to sell its own NFT artwork. *Untitled Magazine.* https://untitled-magazine.com/sophia-the-robot-becomes-first-robot-to-sell-its-own-nft-artwork/

Nussey, S. (2021, June 29). SoftBank shrinks robotics business, stops Pepper production. *Reuters.* www.reuters.com/technology/exclusive-softbank-shrinks-robotics-business-stops-pepper-production-sources-2021-06-28/

Nwagbo, M. A. (2019). *Leadership Trust: A Phenomenological Study of How Major Superiors of Catholic Women Religious Institutes Build Trust with Professed Members* [Doctoral dissertation, Brandman University].

O'Brien, M., & Associated Press. (2023, November 3). Robot startups see huge market in replacing human workers: 'We can sell millions of humanoids, billions maybe'. *Fortune.* https://fortune.com/2023/11/05/when-will-robots-replace-humans-startups-elon-musk-humanoids-optimus/

Odlind, C., & Richardson, K. (2023). The end of sex robots—for the dignity of women and girls. In *Man-Made Women: The Sexual Politics of Sex Dolls and Sex Robots* (pp. 1–16). Cham: Springer International Publishing.

Onyeulo, E. B., & Gandhi, V. (2020). What makes a social robot good at interacting with humans? *Information,* 11(1), 43.

Ornella, A. D. (2015). Uncanny intimacies: Humans and machines in film. In *The Palgrave Handbook of Posthumanism in Film and Television* (pp. 330–338). London: Palgrave Macmillan. https://doi.org/10.1057/9781137430328_33

Ovid, 43 B. C-17 A. D. or 18 A. D. (2004). *Ovid Metamorphoses.* Ed. R. J. Tarrant. Oxford, England: Oxford University Press.

Oz, F. (2004). *The Stepford Wives.* Los Angelos, CA: US. Paramount.

Pagallo, U. (2018). Vital, Sophia, and Co. – The quest for the legal personhood of robots. *Information,* 9(9), 230.

Palmer, C. (2003). *Philip K. Dick: Exhilaration and Terror of the Postmodern.* Liverpool, UK: Liverpool University Press.

Pals, D. L. (2006). *Eight Theories of Religion.* Oxford University Press.

Pandey, K. A., & Gelin, R. (n.d.) A mass-produced sociable humanoid robot: Pepper: The first machine of its kind. *RobotLab Blog.* www.robotlab.com/research-papers/a-mass-produced-sociable-humanoid-robot-pepper-the-first-machine-of-its-kind

Park, T., Goldman, A., Kaplan, C., Mezey, P., Kogonada, Farrell, C., Turner-Smith, J., Min, J. H., & Richardson, H. L. (2022). *After Yang.* Santa Monica, CA: Lionsgate.

Parker, K., & Patten, E. (2013, January 13). *The Sandwich Generation: Rising Financial Burdens for Middle-Aged Americans.* Pew Research Center. www.pewresearch.org/social-trends/2013/01/30/the-sandwich-generation/

Pearce, S. (2013). *On Collecting: An Investigation into Collecting in the European Tradition.* UK: Routledge.

Peterson, R. S. (2016). *The Imago Dei as Human Identity: A Theological Interpretation.* University Park, PA: Penn State University Press.

Philip K Dick Research Robot. (2005). *Hanson Robotics.* www.hansonrobotics.com/philip-k-dick/

Phillips, E., Zhao, X., Ullman, D., & Malle, B. F. (2018). What is humanlike?: Decomposing robots' human-like appearance using the anthropomorphic roBOT (ABOT) Database. In *2018 ACM/IEEE International Conference on Human-Robot Interaction* (pp. 105–113). Chicago: ACM.

Philmus, Robert M. (2001). Matters of translation: Karel Čapek and Paul Selver. *Science Fiction Studies. SF-TH Inc.*, 28(1), 7–32. ISSN 0091–7729. JSTOR 4240948.

Piloto, C. (2023). The gender gap in STEM: Still gaping in 2023. *MIT Professional Education*. https://professionalprograms.mit.edu/blog/leadership/the-gender-gap-in-stem/

Policastro, C. A., Romero, R. A., Zuliani, G., & Pizzolato, E. (2009). Learning of shared attention in sociable robotics. *Journal of Algorithms*, 64(4), 139–151. www.abotdatabase.info/

Prevas, C. (2019, May 9). We need more non-binary characters who aren't aliens, robots, or monsters. *Electric Lit.* https://electricliterature.com/we-need-more-non-binary-characters-who-arent-aliens-robots-or-monsters/

Propp, V. (1968). *Morphology of the Folktale*. Austin, TX: University of Texas Press.

Pu, L., Moyle, W., Jones, C., & Todorovic, M. (2018). The effectiveness of social robots for older adults: A systematic review and meta-analysis of randomized controlled studies. *Gerontologist*, e37–e51. https://doi.org/10.1093/geront/gny046.

Quizzo, E. (2014). How Alderberaan Robotics built its friendly humanoid robot, Pepper. *IEEE Spectrum.* https://spectrum.ieee.org/how-aldebaran-robotics-built-its-friendly-humanoid-robot-pepper

Rabin, N. (2007, January 25). *My Year of Flops, Case File 1: Elizabethtown: The Bataan Death March of Whimsy*. The A.V. Club. Retrieved September 3, 2024: https://www.avclub.com/the-bataan-death-march-of-whimsy-case-file-1-elizabet-1798210595

Realbotix.com. [website]. Retrieved June 16, 2024: https://realbotix.com/

Reeder, H. P. (2010). *The Theory and Practice of Husserl's Phenomenology*. Bucharest, RO: Zeta Books.

Reeves, B., & Nass, C. (1996). *The Media Equation: How People Treat Computers, Television, and New Media Like Real People*. Cambridge: Cambridge University Press.

Richardson, H. (2017, January 30). Robots 'could solve social care crisis,' say academics. *BBC News.* www.bbc.com/news/education-38770516

Richardson, K. (2015). *An Anthropology of Robots and AI: Annihilation Anxiety and Machines*. Milton Park, UK: Routledge.

Richardson, K. (2016). Sex robot matters: Slavery, the prostituted, and the rights of machines. *IEEE Technology and Society Magazine*, 35(2), 46–53.

Richardson, K. (2023). The end of sex robots: Porn robots and representational technologies of women and girls. In *Man-Made Women: The Sexual Politics of Sex Dolls and Sex Robots* (pp. 171–192). Cham: Springer International Publishing.

Ricoeur, P. (1991). Narrative identity. *Philosophy Today*, 35(1), 73.

Riek, L. D., Rabinowitch, T. C., Chakrabarti, B., & Robinson, P. (2009, March 9–13). How anthropomorphism affects empathy toward robots. In *Proceedings of the 4th ACM/IEEE International Conference on Human Robot Interaction* (pp. 245–246). LaJolla, CA and New York, NY: ACM/IEEE. Retrieved September 4, 2024: https://ieeexplore.ieee.org/xpl/conhome/1040036/all-proceedings

Robertson, J. (2019). *Robo Sapiens Japanicus: Robots, Gender, Family, and the Japanese Nation*. Oakland, CA: University of California Press.

Robinson, N. L., Connolly, J., Suddrey, G., & Kavanagh, D. J. (2023, October 12). A brief wellbeing training session delivered by a humanoid social robot: A pilot randomized controlled trial. *International Journal of Social Robotics*, 1–15. https://doi.org/10.1007/s12369-023-01054-5

Robots Guide. (2024). *Geminoid HI-1*. Retrieved September 4, 2024: https://robotsguide.com/robots/geminoidhi1

Robots Guide. *Jibo*. Retrieved September 5, 2024: https://robotsguide.com/robots/jibo

Rosenthal-Von Der Pütten, A. M., Schulte, F. P., Eimler, S. C., Sobieraj, S., Hoffmann, L., Maderwald, S., Brand, M., & Krämer, N. C. (2014). Investigations on empathy towards humans and robots using fMRI. *Computers in Human Behavior*, 33, 201–212.

Roy, A. (2021). What we can learn from the era of ASIMO: The future of robotics. *Medium*. https://bootcamp.uxdesign.cc/what-we-can-learn-from-the-era-of-asimo-the-future-of-robotics-9bc0b03abc45

Roy Batty. (n.d.). *Off-World: The Blade Runner Wiki*. Fandom, Inc. https://blade-runner.fandom.com/wiki/Roy_Batty_(templant)

Russell, S., Dewey, D., & Tegmark, M. (2015). Research priorities for robust and beneficial artificial intelligence. *AI Magazine*, 36(4), 105–114. Washington, DC: AAAI.

Saerbeck, M., & Bartneck, C. (2010, March). Perception of affect elicited by robot motion. In *2010 5th ACM/IEEE International Conference on Human-Robot Interaction (HRI)* (pp. 53–60). IEEE.

Sakagami, Y., Watanabe, R., Aoyama, C., Matsunaga, S., Higaki, N., & Fujimura, K. (2002, September). The intelligent ASIMO: System overview and integration. In *IEEE/RSJ International Conference on Intelligent Robots and Systems* (Vol. 3, pp. 2478–2483). IEEE.

Salichs, M. A., Encinar, I. P., Salichs, E., Castro-González, Á., & Malfaz, M. (2016). Study of scenarios and technical requirements of a social assistive robot for Alzheimer's disease patients and their caregivers. *International Journal of Social Robotics*, 8, 85–102.

Saari, U. A., Tossavainen, A., Kaipainen, K., & Mäkinen, S. J. (2022). Exploring factors influencing the acceptance of social robots among early adopters and mass market representatives. *Robotics and Autonomous Systems*, 151, 104033.

Samuel, S. (2020, January 13). Robot priests can bless you, advise you, and even perform your funeral. *Vox*. Retrieved May 2, 2024: www.vox.com/future-perfect/2019/9/9/20851753/ai-religion-robot-priest-mindar-buddhism-christianity

Sarmah, T. (2006). *New Trends in Interpretation of the Vedas*. New Delhi: Sundeep Prakashan.

Sartre, J. P. (2015). *Being and Nothingness. Central Works of Philosophy v4: Twentieth Century: Moore to Popper* (Vol. 4, p. 155). Milton Park, UK: Routledge.

Saudi Arabia: Law Enshrines Male Guardianship. (2023, March 8). Human Rights Watch. www.hrw.org/news/2023/03/08/saudi-arabia-law-enshrines-male-guardianship

Saunderson, S. P., & Nejat, G. (2015). Persuasive robots should avoid authority: The effects of formal and real authority on persuasion in human-robot interaction. *Autonomous Robots*, 39, 293–312.

Schubumkehr Manifesto, 1996. (2021). "Manifiestos sobre el arte y la red 1990–1999." *Exit Media*. Retrieved 25 March 2021.

Schlosser, L. (2003). Christian Privilege: Breaking a sacred taboo. *Journal of Multicultural Counseling and Development*, 31(1), 44–51.

Schneider, A. F., Grenzfurthner, J., & Friesinger, G. (2024). Context hacking: How to mess with art, media, law, and the market. *Monochrome*. www.monochrom. at/context-hacking/

Schreier, J. (2012). *Robot & Frank*. Culver City, CA: Sony Pictures Home Entertainment.

Scott, R. (1982, June 25). *Blade Runner*. Burbank, CA, USA: Warner Brothers.

Searle, J. R. (1980). Minds, brains, and programs. *Behavioral and Brain Sciences*, 3(3), 417–424.

Seelman, K. D. (2016). Should robots be personal assistants? In *Occupying Disability: Critical Approaches to Community, Justice, and Decolonizing Disability* (pp. 259–272). Dordrecht: Springer. https://doi.org/10.1007/978-94-017-9984-3_18

Segall, L. (2017). Mostly human: I love you, bot. *CNN*. https://money.cnn.com/mostly-human/i-love-you-bot/

Seifert, T. (2007). Understanding Christian privilege: Managing the tensions of spiritual plurality. *About Campus*, 12(2), 10–17.

Semuels, A. (2022, August 6). Millions of Americans have lost jobs in the pandemic, and robots and AI are replacing them faster than ever. *TIME*. https://time.com/5876604/machines-jobs-coronavirus/

Serpell, J. (1996). *In the Company of Animals: A Study of Human-animal Relationships*. Cambridge University Press.

Sex Toys: Global Strategic Business Report. (2024). [Report]. Research and Markets. San Jose, CA: Global Industry Analysts, Inc. ID: 5140296. www.researchand-markets.com/reports/5140296/sex-toys-global-strategic-business-report

Sharkey, A., & Sharkey, N. (2012). Granny and the robots: Ethical issues in robot care for the elderly. *Ethics and Information Technology*, 14, 27–40. Springer.

Sharkey, N., & Sharkey, A. (2020). *The Crying Shame of Robot Nannies: An Ethical Appraisal. In Machine Ethics and Robot Ethics* (pp. 155–184). London: Routledge.

Shaw, D. B. (1994). *The Power of Assumptions and the Power of Poetry: A Reading of Ovid's "Tristia 4"*. Berkeley: University of California.

Shibata, T. (2012). Therapeutic seal robot as biofeedback medical device: Qualitative and quantitative evaluations of robot therapy in dementia care. *Proceedings of the IEEE*, 100(8), 2527–2538. https://doi.org/10.1109/EMBC.2012.6346772

Shibata, T., Hung, L., Petersen, S., Darling, K., Inoue, K., Martyn, K., Hori, Y, Lane, G., Park, D., Mizoguchi, R., Takano, C., Harper, S., Leeson, G. W., & Coughlin, J. F. (2021). PARO as a biofeedback medical device for mental health in the COVID-19 Era. *Sustainability*, 13(20), 11502. https://doi.org/10.3390/su132011502

Shigemi, S. (2018, October 10). ASIMO and humanoid robot research at Honda. In A. Goswami & P. Vadakkepat (Eds.), *Humanoid Robotics: A Reference*. Dordrecht: Springer. https://doi.org/10.1007/978-94-007-6046-2_9

Shirow, M. (1989-1991). *Ghost in the Shell*. Tokyo: Kodansha.

Siegel, M., Breazeal, C., & Norton, M. I. (2009, October). Persuasive robotics: The influence of robot gender on human behavior. In *2009 IEEE/RSJ International Conference on Intelligent Robots and Systems* (pp. 2563–2568). IEEE.

Singularitynet.com. (2024). *About Us*. https://singularitynet.io/aboutus/

Sini, R. (2017, October 26). Does Saudi robot citizen have more rights than women? *BBC*. Retrieved August 24, 2024: https://www.bbc.com/news/blogs-trending-41761856

Smith, J. K. (2021). *Robotic Persons: Our Future with Social Robots*. New York: Westbow Press.

Smith, J. Z. (1987). *To Take Place: Toward Theory in Ritual*. Chicago, IL: University of Chicago Press.

SoftBank Mobile and Aldebaran unveil Pepper—the world's first personal robot that reads emotions. (2014, June 5). N.A. SoftBank Mobile Corp. www.softbank.jp/en/corp/group/sbm/news/press/2014/20140605_01/

Sohrabi-Shiraz, A. (2018, January 9). Sex robot maker Sergi Santos: Who is he and what does Synthea Amatus do? *Daily Star*. www.dailystar.co.uk/news/latest-news/sex-robot-sergi-santos-who-16837966

Song, Y., Luximon, A., & Luximon, Y. (2021). The effect of facial features on facial anthropomorphic trustworthiness in social robots. *Applied Ergonomics*, 94, 103420.

Sparrow, R. (2020a). Robotics has a race problem. *Science, Technology, & Human Values*, 45(3), 538–560.

Sparrow, R. (2020b, November 27). Sex robot fantasies. *Journal of Medical Ethics*. https://doi.org/10.1136/medethics-2020-106932.

Sparrow, R., & Sparrow, L. (2006). In the hands of machines? The future of aged care. *Minds and Machines*, 16, 141–161.

Spielberg, S. (2001). *Artificial Intelligence: AI*. [film]. US: Warner Brothers.

Spitzer, K. (2015, September 29). Warning: Please don't have sex with the robot. *USA Today*. www.usatoday.com/story/news/world/2015/09/29/warning-please-dont-have-sex-robot/73020322/

Sroufe, L. A., Carlson, E., & Shulman, S. (1993). Individuals in relationships: Development from infancy through adolescence. In D. C. Funder, R. D. Parke, C. Tomlinson-Keasey, & K. Widaman (Eds.), *Studying Lives through Time: Personality and Development* (pp. 315–342). Washington, DC: American Psychological Association. https://doi.org/10.1037/10127-030

Stano, S. (2024). Dis-and re-embodiment in religious practices: Semiotic, ethical, and normative implications of robotic officiants. *International Journal for the Semiotics of Law-Revue internationale de Sémiotique juridique*, 37(4), 1209–1221.

Stake, R. E. (1995). *The Art of Case Study Research*. Thousand Oaks, CA: Sage.

Stark, M. (2021, April 17). *Rachael Ma: Performer, Choreographer and Teacher— Theatre People*. www.theatreartlife.com/lifestyle/rachael-ma-performer-choreographer-and-teacher-theatre-people/

Suchman, L. A. (1987). *Plans and Situated Actions: The Problem of Human-Machine Communication.* Cambridge: Cambridge University Press.

Sullins, J. P. (2012). Robots, love, and sex: The ethics of building a love machine. *IEEE Transactions on Affective Computing,* 3(4), 398–409.

Sung, J. Y., Guo, L., Grinter, R. E., & Christensen, H. I. (2007). "My Roomba is Rambo": Intimate home appliances. In *UbiComp 2007: Ubiquitous Computing: 9th International Conference, UbiComp 2007, Innsbruck, Austria, September 16–19, 2007. Proceedings 9* (pp. 145–162). Berlin and Heidelberg: Springer.

Sutin, L. (1995). *The Shifting Realities of Philip K. Dick: Selected Literary and Philosophical Writings* (p. 212). New York: Pantheon Books.

Swackhamer & Russell, 1961–1966. *Hazel* [Television]. NBC: 1961–65; CBS: 1965–66.

Sweeney, P. (2023). Trusting social robots. *AI and Ethics,* 3(2), 419–426.

Swift, C. (2015). *Preternature: Critical and Historical Studies on the Preternatural* (Vol. 4, No. 1, Special Issue: Animating Medieval Art, pp. 52–77). Penn State University Press. www.jstor.org/stable/10.5325/preternature.4.1.0052

Symchuk, A. (2024, February 10). Johannes Grenzfurthner: The Austrian director redefining horror. *MovieWeb.* Retrieved April 30, 2024: https://movieweb.com/johannes-grenzfurthner-director-redefining-horror/

Syrdal, D. S., Walters, M. L., Otero, N., Koay, K. L., & Dautenhahn, K. (2007). He knows when you are sleeping-privacy and the personal robot companion. In *Proc. Workshop Human Implications of Human-Robot Interaction, Association for the Advancement of Artificial Intelligence (aaai'07)* (pp. 28–33). Washington, DC: AAAI.

Talmud, The. (2024). The William Davidson Edition. *Sefaria.org.* New York, NY. Retrieved September 4, 2024: https://www.sefaria.org/texts/Talmud

Tarantola, A. (2016, April 1). Inventor builds super creepy 'Scarlett Johansson' robot. *Endgadget.* www.engadget.com/2016-04-01-inventor-builds-super-creepy-scarlett-johansson-robot.html

Teo, Y. (2020). Recognition, collaboration and community: Science fiction representations of robot carers in Robot & Frank, Big Hero 6 and Humans. *Medical Humanities,* 47, 95–102.

Terasem Faith. (2024). *Join.* Retrieved August 26, 2024: https://terasemfaith.net/login-2/

Terasem Movement Foundation. (2024). *About TMF.* Retrieved August 26, 2024: https://terasemmovementfoundation.com/

Tezuka Osama Official. *Astro Boy.* Retrieved October 4, 2024: https://tezukaosamu.net/en/manga/291.html

Tezuka, O. (1952–1968). *Astro Boy* [Manga series]. Kobunsha: Shōnen.

Tezuka, O. (2015, October 13). *Astro Boy Omnibus* (Vol. 1). Milwaukie, OR: Dark Horse Manga [Illustrated edition].

Thalman, D. (1996). The complexity of simulating virtual humans. *Supercomputing Review,* 8, 16–23.

Thalmann, N. M., & Thalmann, D. (1987a). *Rendez-vous in Montreal* [Film]. University of Montreal.

Thalmann, N. M., & Thalmann, D. (1987b). The direction of synthetic actors in the film Rendez-vous in Montreal. *IEEE Computer Graphics and Applications,* 7(12), 9–19.

Todd, R. (2015, September 17). There might not have been an iRobot without Rosie the Robot: How a fascination with The Jetsons's maid inspired a whole generation of roboticists, including iRobot CEO and cofounder Colin Angle. FastCompany.www.fastcompany.com/3051214/there-might-not-have-been-an-irobot-without-rosie-the-robot

Totodigio. (2008, January 9). Hajime Sorayama: The airbrush legend. *Total Graphic.* https://totalgraphic.wordpress.com/maestros/hajime-sorayama/

Toyota Blog. (2015, February 12). *Mission Accomplished: Robot Astronaut Kirobo Returns to Earth.* https://mag.toyota.co.uk/kirobo-space-robot-return-earth/

Trung, L. (2024). *Project Aiko.* Retrieved August 22, 2024: http://projectaiko.com/

Turkle, S. (2011). *Alone Together: Why We Expect More from Technology and Less from Each Other.* New York: Basic Books.

Turkle, S. (2020). A nascent robotics culture: New complicities for companionship. In *Machine Ethics and Robot Ethics* (pp. 107–116). London: Routledge.

Turner, J. (2018). *Robot Rules: Regulating Artificial Intelligence.* New York: Springer.

Turney, D. (2022, August 12). The history of industrial robots, from single task-master to self-teacher. *Autodesk.* www.autodesk.com/design-make/videos/history-of-industrial-robots

Ulanoff, L. (2003, January 28). ASIMO Robot to tour U.S.A. *PC Magazine.* Archived from the original on 14 October 2012. https://web.archive.org/web/20121014091718/www.pcmag.com/article2/0,2817,849588,00.asp

Underwood, D. W. (2013). *American Socio-Politics in Fictional Context: Transformers and the Representation of the United States* [Doctoral dissertation, University of East Anglia].

United Nations Department of Economic and Social Affairs, Population Division (2019). Retrieved October 4, 2024: https://www.un.org/development/desa/pd/

United Robotics Group. (2015). *Human Emotions Decoded, Embraced, Echoed: Pepper.* https://unitedrobotics.group/en/robots/pepper

U.S. Department of State. (2022). *2022 Report on International Religious Freedom: Saudi Arabia.* [Report]. www.state.gov/reports/2022-report-on-international-religious-freedom/saudi-arabia

Vandemeulebroucke, T., de Casterlé, B. D., & Gastmans, C. (2018). How do older adults experience and perceive socially assistive robots in aged care: a systematic review of qualitative evidence. *Aging & Mental Health*, 22(2), 149–167.

Verstegen, D. (2017, September 14). Someone invented an artificially intelligent sex robot. *USA Today.* www.usatoday.com/story/life/allthemoms/news/2017/09/14/samantha-sex-robot/34915203/

Villaneuve, D. (2017, October 3). *Blade Runner: 2049.* [Movie]. North America: Warner Brothers.

Volkmann, K. (2011, April 28). Honda's ASIMO visits FIRST robotics event. *St. Louis Business Journal.* www.bizjournals.com/stlouis/blog/2011/04/hondas-asimo-visits-first-robotics.html

Von Harbou, T. (2011, June 8). *Metropolis.* Kindle: Amazon.

Vygotsky, L. S. (1978). *Mind in Society.* Cambridge, MA: Harvard University Press.

Vygotsky, L. S. (1986). *Thought and Language*. Transl. and ed. A. Kozulin. Cambridge, MA: MIT Press (Originally pub. in 1934).

Vygotsky, L. S. (2004). Imagination and creativity in childhood. *Journal of Russian and East European Psychology*, 42(1), 7–97.

Wada, K., & Shibata, T. (2007). Living with seal robots—its sociopsychological and physiological influences on the elderly at a care house. *IEEE Transactions on Robotics*, 23(5), 972–980.

Walters, H. (2023, March). Robots are performing Hindu rituals- some devotees fear they'll replace worshippers. *The Conversation*. https://theconversation.com/robots-are-performing-hindu-rituals-some-devotees-fear-theyll-replace-worshippers-197504

Watts, J. (26 January, 2019). *Sexism by design: Gender representation in roboics is setting women up for a regressive future*. San Francisco, CA: Medium.

Weiss, J. (2022, October 13). *Clown, the Gonzo Body Horror Film, Started with a Fake Trailer*. Syfy Wire. Retrieved September 4, 2024: https://www.syfy.com/syfy-wire/clown-horror-movie-christopher-ford-screenwriter-interview

Wendig, C. (2016). *Star Wars: Aftermath*. New York: Del Ray. [Kindle].

Wendig, C. (2019). *Wanderers*. New York: Del Ray.

Wertsch, J. V. (1991). *A Sociocultural Approach to Socially Shared Cognition*. American Psychological Association.

White, C. (2007, August 9). Honda ASIMO to Return to Disneyland. *Gizmodo*. https://gizmodo.com/honda-asimo-to-return-to-disneyland-run-around-and-tal-294650

Winfield, A. (2011). Roboethics—for humans. *New Scientist*, 210(2811), 32–33.

Winnicott, D. W. (1989). *Playing and Reality*. London: Routledge.

Wolf, R. (2020, September 11). Premee Mohamed is not a 'new' writer. *LitHub*. https://lithub.com/premee-mohamed-is-not-a-new-writer/

Wortham, J. (2012, August 10). From the future, a subtle spark of recognition. *New York Times*. www.nytimes.com/2012/08/12/movies/in-robot-frank-technology-of-the-not-so-distant-future-on-display.html

Wright, J. (2023, January 9). Inside Japan's long experiment in automating elder care. *Technology Review*. www.technologyreview.com/2023/01/09/1065135/japan-automating-eldercare-robots/

Xuan, Y. L. (2024, March 8). Video of Saudi Arabia's first male robot goes viral for the wrong reason. *The Straits Times*. www.straitstimes.com/world/middle-east/video-of-saudi-arabia-s-first-male-robot-goes-viral-for-the-wrong-reason

Yin, R. K. (2017). *Case Study Research and Applications: Design and Methods*. Thousand Oaks, CA: Sage Publications.

Zawieska, K., Duffy, B. R., & Strońska, A. (2012). Understanding anthropomorphisation in social robotics. *Pomiary Automatyka Robotyka*, 16(11), 78–82.

Zitser, J. (2024, March 11). A male humanoid robot inappropriately touches a female reporter. *Business Insider*. www.businessinsider.com/saudi-male-humanoid-robot-inappropriately-touched-female-reporter-video-shows-2024-3

Index

Note: Page numbers in *italic* indicate a figure on the corresponding page.